The Language of Popular Science

The Language of Popular Science

Analyzing the Communication of Advanced Ideas to Lay Readers

OLGA A. PILKINGTON

McFarland & Company, Inc., Publishers
Jefferson, North Carolina

LIBRARY OF CONGRESS CATALOGUING-IN-PUBLICATION DATA

Names: Pilkington, Olga A., author.
Title: The language of popular science : analyzing the communication of advanced ideas to lay readers / Olga A. Pilkington.
Description: Jefferson, N.C. : McFarland & Company, Inc., Publishers, 2019 | Includes bibliographical references and index.
Identifiers: LCCN 2018054289 | ISBN 9781476672533 (softcover : acid free paper) ∞
Subjects: LCSH: Science—Language. | Science—Popular works. | Discourse analysis. | English language—Discourse analysis.
Classification: LCC Q223 .P548 2019 | DDC 501/.4—dc23
LC record available at https://lccn.loc.gov/2018054289

BRITISH LIBRARY CATALOGUING DATA ARE AVAILABLE

ISBN (print) 978-1-4766-7253-3
ISBN (ebook) 978-1-4766-3560-6

© 2019 Olga A. Pilkington. All rights reserved

No part of this book may be reproduced or transmitted in any form or by any means, electronic or mechanical, including photocopying or recording, or by any information storage and retrieval system, without permission in writing from the publisher.

Front cover images: (middle) model of particle collisions in the Large Hadron Collider (General-FMV/Shutterstock); (bottom) X-rays streaming off the sun in an image from the Nuclear Spectroscopic Telescope Array, or NuSTAR (NASA)

Printed in the United States of America

McFarland & Company, Inc., Publishers
  Box 611, Jefferson, North Carolina 28640
    www.mcfarlandpub.com

To Ace, Alexander
and Natalie

Acknowledgments

There are many people who deserve my gratitude in connection with this book. First of all, I wish to thank Dr. Ace G. Pilkington, my husband, for his encouragement, enthusiasm, and continuous help with the writing. I am also grateful to Professor Susan E. Hunston, OBE (University of Birmingham, UK), who read and commented on the earlier versions of many of these chapters.

My thanks also go to the library staff of Dixie State University and the interlibrary loan department of the Brigham Young University Library for their prompt delivery of hard-to-find sources.

Of course, this book would not be possible without the help of McFarland's staff. I am especially grateful to Robert Franklin, who suggested the idea for this book, and to Charles L. Perdue for guiding me through the publishing process.

Finally, I want to say thank you to Alexander and Natalie Ivanchenko, my parents, whose support always means a great deal to me.

Table of Contents

Acknowledgments   vi
Preface   1
Introduction: Popular Science   4

1. A Linguist Looks at Popular Science   17
2. Personal Narratives   44
3. Narratives of Discovery: Explanation Made Easy   59
4. Narratives and Ideology: What's in a Structure?   70
5. What They Say: Speech of Scientists   91
6. What They Imagine Is Possible: Thoughts of Scientists   106
7. Literature and Limericks: Writing in Popular Science   115
8. Definitions: Types and Methods   122
9. Interacting with Readers through Definition   135
10. The Fictionalized Reader   150
11. Lab Lit: Fictional Science   156

Conclusion: Professional Science and Popular Science   164
Epilogue   172
Sources   179
Index   187

Preface

It all began with Stephen Hawking. As Marcus Chown, author of *We Need to Talk About Kelvin*, explains, "There have been popular science writers before—Carl Sagan, Isaac Asimov. But I don't believe there were popular science sections in bookshops before Hawking" (cited in Chivers 2010). Max Genecov (2018), writing just a month after Hawking's death, describes *A Brief History of Time* as "one of the first books meant for a general audience about cosmology." Today, any major bookstore has a popular science section, and scientists are actively encouraged to popularize their research and to share the findings with the lay public as well as with colleagues. In fact, being a popularizer *and* a working scientist has become a badge of honor so much so that Genecov chose to open his tribute to Stephen Hawking with a mention of his public engagement: "Stephen Hawking … was a physicist from another time. He had more in common with … the politically-inclined scientist-intellectuals like Robert Oppenheimer, Richard Feynman, and Enrico Fermi who came out of the Manhattan Project—than he did with any of his contemporaries. This isn't so much a question of brainpower as it is of public positioning, as getting the public to understand new scientific ideas is a very different job than coming up with them."

Yet, combining these two jobs is what the modern public demands. "Universities … are increasingly being asked to get their academics and other staff out of their ivory silos to talk to and listen to the various publics, individuals and organizations in the communities they are embedded in … and the pressure is increasing on grant holders to engage more effectively with the public" (Roberts 2013: xix). That means learning to write in a way that will appeal to a non-specialist. Multiple writing manuals teaching scientists to popularize effectively have been published alongside popular science books themselves. Popular science is as popular as ever!

New names have emerged since Asimov, Clarke, Gamow, Sagan and Hawking. Today's laymen are reading Bill Bryson, Brian Greene and Michio Kaku, among others. But what makes their books so popular? What can their writings reveal to an emerging popularizer?

It turns out that popular scientific prose is quite homogenous in the way it approaches its scientific subject matter. For example, Turney (2007: 2) observed that certain metaphors and analogies were repeated in popular science so often that they formed a pool of stock imagery that many authors "adopt and modify" for their own purposes. In contrast, new metaphors, according to Turney (2004b: 337), indicate that "there is not yet a widely accepted formula for describing … novel" ideas. Turney (2004b: 343) suggests that the adaptation and modification of certain metaphors indicate the success of their originator.

The similarities among popular science works go deeper than formulaic language. They are structural. In this book, I will focus on three linguistic features I have identified as used similarly in a variety of popular science books. They include narratives, definitions of scientific terminology, and the way scientists' speech, thoughts, and sometimes writing are reflected in popularizations (that is, presented discourse of scientists).

I begin with a brief excursion into the history of popular science and tie it in with the modern state of the genre in the introduction. Chapter 1 gives background and general explanations of the three features of popular science that make up the main subject of this book. The subsequent chapters discuss storytelling, definitions, voices of scientists, and the relationship between readers and writers. Each chapter introduces ample details and includes examples from Marcus du Sautoy, Timothy Ferris, and Sam Kean among other well-known popular science authors. The concluding chapters of the book place popular science in a continuum of scientific writing that includes professional research articles and fiction about scientists and the laboratory. Here I draw on lab lit novels by Susan Gaines, Allegra Goodman, Alan Lightman, and Jennifer Rohn.

While this work began as a PhD project at the University of Birmingham, UK, in its present form it is designed for a general audience. I especially hope that this book will find its way into the hands of philosophers, historians, and sociologists of science since it adds a new perspective (that of a linguist) to the existing discussions of the role and importance of popular science. When possible, I avoid the use of linguistics jargon, and when it is necessary to rely on the discipline-specific vocabulary, I include explanations. I envision this book as a kind of guide into the world of popular science writing for both the readers and the writers of popularizations. This is not a manual but rather an analytical journey through popular science books written by famous authors in order to see what makes this literature successful. For a popular science enthusiast, it may shine new light on the beloved texts and reveal them not only as sources of scientific information but also as examples of the scientific community's attempts to appear benevolent, interesting, and appealing to the general public. For a science writer, this work may reinforce the importance of certain techniques and perhaps even suggest a few new ones.

Though I grew up in a family of chemists, I never had any particular interest in science. This project, in many ways, is my personal journey through the world I was born into but never truly embraced. For me, popular science traced the path through the scientific wilderness that I did not think I could traverse. As a result, I wanted to know what made these texts so engaging and relatable. In this book, I invite you to re-trace my steps to an understanding of popular science beyond the scientific facts.

Introduction: Popular Science

The power and influence of modern science is undeniable. So is the desire to be "in the know," to understand the how and the why of the universe. The quest for knowledge belongs, however, not only to the professional scientist but to the lay person as well. Despite multiple surveys that show how remarkably unaware of certain scientific matters the public is, the general interest in science among non-professionals remains strong. And as Myers (2003: 268) shows, "nor is the public entirely cut off from expertise." Lay people tend to know the kind of science that reflects certain personal interests or allows for an emotional connection—"more people are interested in health and risk issues than in, say, algebraic theory or materials science" (Myers 2003: 269). For people who are interested in science but reside outside the professional scientific community, popular science offers the answers—literally.

It would not be unusual to assume that the line between professional science and popularizations is a straight one—from one kind of explanation to another, simpler and without the math. Popular science, though, is not just a matter of interpretation. It is a complex linguistic and ideological construct that often operates under the same rules and regulations as the scientific community.

The idea that professional science is for the educated elite and popular science is for the masses simply does not withstand a careful investigation. From the time when science emerged as a discipline to the latest popular article posted on the internet, there has been an overlap between the specialist and the spectator. In fact, the idea of professional science as opposed to amateur interest did not emerge until the 19th century and was the result of deliberate actions by a few members of the scientific elite.

Up until then, there are examples of active public involvement and of science being accessible to lay persons. This kind of "open-door" approach attracted many amateurs who ended up making scientific contributions (see, for example, Bensaude-Vincent 2001; Lightman 2000; Topham 2000). Today these people are called "enlightened amateurs" (Bensaude-Vincent 2001: 102). However, at the time, they "considered themselves ... members of the republic of science, a large international community ... of people who investigated nature and reported their results to each other" (Bensaude-Vincent 2001: 102). From the second half of the 19th century, however, science shut its doors to anyone who was not a professional, and that is how the famous "gap" between the scientists and the public came to be.

Some believe that the gap was a result of deliberate and detrimental actions of a small group of scientists led by T.H. Huxley (see, for example, Lightman 2000). Their actions became known as boundary work—a term that is still used today to describe focused efforts to discourage public participation and influence in scientific matters. As Lightman (2000: 101) sees it, scientists of the 1890s tried to secularize science by separating it from what he calls "elements that previously had connected public and scientific culture, including anthropomorphic, anthropocentric, teleological and ethical views of nature." The professional scientists of the second half of the 19th century set out to establish themselves as the only authorities on secular, natural knowledge. As such, they favored "detached language, heavily spiced with complex scientific terms" designed to marginalize and eventually push out the non-experts (Lightman 2000: 101). Such use of language deliberately expunged story telling from professional scientific writing, making scientific results appear as entirely objective and untainted by human interference.

Another factor that contributed to the separation of professional science from the public was the dramatic change in the complexity of scientific procedures and the new apparatus involved that now required certain specific kinds of expertise rather than general experience and curiosity. Topham (2000: 560) calls this process a series of "epistemological and rhetorical shifts" that positioned scientific discoveries "as the preserve of scientific 'genius'" rather than as an activity open to anyone.

Such elevation of the status of professional science, created a very peculiar model for popularizations. As the public interest in science never really diminished, popularizers had to find a way to deliver scientific knowledge in a manner that would not undermine the superior position of the professional scientific community. Enter what became known as "familiar science" (Keene 2014).

The main underlying assumption behind the notion of familiar science was that the explanation to laymen would progress more smoothly if a popularizer referred almost exclusively to the objects, contexts and environments already well-known to the public. This kind of popularization drew heavily on domestic objects and local knowledge, encouraging people to engage with science, but only up to a point. In effect, instead of opening the doors of the laboratory, familiar science replaced it with the dinner table and the backyard. Keene provides an excellent illustration of this process by analyzing T.H. Huxley's 1868 lecture "On a Piece of Chalk" (2014: 64–65). She writes, "when discussing the chemical composition of chalk," Huxley uses an example of a simple object easily found in any kitchen—a tea kettle. "The fur on the inside of a tea-kettle," Huxley explains, "is carbonate of lime; and, for anything chemistry tells us to the contrary, the chalk might be a kind of gigantic fur upon the bottom of the earth-kettle, which is kept pretty hot below" (cited in Keene 2014: 65). In Keene's (2014: 65) words, "Huxley's listeners were encouraged to conflate their domestic world with the world of nature." They were not to examine (even if only by the power of their imaginations) the properties of chalk as a scientist would—through laboratory procedures involving apparatus—but as true laymen and outsiders whose only possible point of reference was the home and the objects found in it.

Even though as time went on the familiar science approach was replaced by more sophisticated means of communicating science to the public, examples of this kind of popularization ideology exist as late as 1984. In *The Limits of Science*, Peter Brian Medawar proposes a generally noble idea that the basic principles of science are accessible to an average lay person. The example he gives, however, evokes a Victorian-era approach as he chooses to illustrate his idea with an image of a housewife repairing a lamp.

There is nothing wrong in explaining scientific concepts that may appear alien to some readers through allusions to environments and objects familiar to them. It is wrong, though, to suggest that the only kind of science available to a non-professional is a cross between magic tricks and cooking disasters. Using language and imagery that a lay person is likely to respond to does not automatically exclude intelligent and provocative discussions that employ scientific terminology, describe and analyze actual laboratory activities, and talk the readers through thought experiments akin to those scientists perform.

Modern popular science does just that. However, it is more sophisticated not only for the sake of the reader but for its own, hidden purposes.

The notion that contemporary popular science (be it books or blogs or articles in *Scientific American*) is written exclusively for consumption by the lay public is flat out wrong. Popular science, it turns out, has a very professional audience. As Myers (2003) points out, even a well-educated scientist cannot be a specialist in all fields and branches even of her own discipline let alone the multitude of areas of knowledge that today's science covers. Myers (2003: 268) explains, "When I go to the doctor, I treat her as an expert in medicine, but her relation to current medical research will generally be as a continuing student, not as a participant, and the medical journals have to perform a kind of popularizing function for her." But it is not only the traditional medical journals that a modern doctor is likely to consult. A study published in *The New England Journal of Medicine* in 1991 showed "that articles appearing in *The New England Journal of Medicine* are cited twice as much by specialists if they are also mentioned in a daily paper like the *New York Times*" (Bucchi 1998: 11). Myers (2003: 268) correctly acknowledges that along with his physician, "Administrators, medical students, patent lawyers, post-docs, technicians, science journalists, and research scientists in commercial firms all take on, sometimes uncomfortably, an identity between expert and lay." And that is one of the reasons for modern popular science being different from popularizations of the past.

Thus, contemporary popular science serves as a communication channel not only between the scientific community and the layman but also among the professionals within the laboratory walls. As such, popular science has become a forum for introducing new ideas and denouncing the competition—a function previously reserved for professional peer-reviewed publications. "Scientific discourse at the public level is only apparently 'public': communication at this level is not actually meant to address the general public, but to reach a vast number of colleagues rapidly … without having to conform to the times and constraints of specialist communication" (Bucchi 1998: 12). Popular science outlets are favored by professional scientists because they allow for a much easier dissemination of research findings than any professional publication can. Bypassing the rigorous peer-review stage of the publication process, a popularized account can appear in a magazine or a newspaper in a matter of hours, not months—a typical timeframe for academic research articles. Bucchi (1998: 12) proposes another possible reason for using a popular instead of a professional route to publication: popular science essentially strips research findings from the very complex contexts in which they were obtained and allows for an easier and more effective interpretation that could be bent to present the results as more authoritative than they actually are.

I should note at this point, that popular science often creates its own context for the discoveries it describes—this is usually achieved through the use of narratives. As Reitsma (2010: 93) notes, "A narrative ... includes interpreted information," which, in turn, eliminates the need for the reader to draw his own connections among the events presented. While this may seem to be a form of knowledge control, let's not forget that a reader who gets information from popularized accounts is very likely not equipped with enough background information to reach logical conclusions on her own. Turney (2004b: 333) surveys the kinds of narratives employed in popular science and suggests that they all serve one purpose—to explain or "translate" science into laymen's terms. Recent studies (see, for example, Blanchard et al. 2015; Hermwille 2016; Reitsma 2010) demonstrate that the explanatory and contextualizing abilities of popular science narratives appeal not only to the science-minded laymen but also to the decision-making social power structures such as grant-providing agencies or policy-creating institutions—in other words to the semi-professionals. Here is a description of a fairly recent discovery of viruses in an environment not contaminated by any living organisms:

> In 2009, another scientist, Curtis Suttle, paid a visit to the Cave of Crystals. Suttle and his colleagues scooped up water from the chamber's pools and brought it back to their laboratory at the University of British Columbia to analyze. When you consider Suttle's line of work, his journey might seem like a fool's errand. Suttle has no professional interest in crystals, or minerals, or any rocks at all for that matter. He studies viruses. There are no people in the Cave of Crystals for the viruses to infect. There are not even any fish. The cave has been effectively cut off from the biology of the outside world for millions of years. Yet Suttle's trip was well worth the effort. After he prepared his samples of crystal water, he put them under a microscope and saw protein shells loaded with genes. Each drop of cave water may hold two hundred million viruses [Zimmer 2011: 2].

This narrative will easily appeal to any non-specialist, whether a doctor, a poet, a physicist or a construction worker. It is simple, yet it does not shy away from scientific terminology ("protein shells"), mentions of the laboratory, and its equipment ("a microscope"). It also presupposes that a reader is familiar with the basic ideas about viruses: what they are and how they originate and spread. Otherwise, the references to the absence of people "and even any fish" in the cave would make very little sense.

One of the implications of a dual—lay and professional—audience for popular science is the debates and disagreements that play out on the pages of popular science books. Turney (2007: 2) supplies an excellent example: "Some of the physicists' arguments—shorn of the maths—are being fought out in popular books. Proponents of superstring theory such as Brian Greene are under attack by those who favour other approaches like Lee Smolin partly because they have been so good at promoting their ideas to the wider public. Smolin's recent *The Trouble with Physics* is an attack on PR as much as on superstring theory."

Making popular science essentially into a forum for professional scientific debate brought the complexity and the sophistication of the discussions back to the lay reader. At the same time, this literature is still written primarily for a non-specialist audience who do not possess any kind of scientific training. In order to appeal to both potential readerships, popular science has developed specific mechanisms for presenting information that might be hard to understand for the public. These include well-structured and prolonged definitions of scientific terminology (see chapter 8), narrative technique (see chapters 2–4), and the use of the voices of scientists to fictionalize material and appeal to readers' emotions (see chapters 5–7).

In fact, emotional connection with professional science has become the path back to the laboratory for a lay enthusiast. However, because of the increasing complexity of scientific activities and the apparatus involved, it is nearly impossible for an ordinary individual to contribute to the production of scientific knowledge in the same way that was feasible in the past. For the modern lay public, engagement with science shows up in the emotional engagement with and the evaluation of newly discovered scientific realities. While many scientists object to this involvement, the public has the power to declare some research procedures unethical and, through political representatives, cut off federal or local government funds for a project.

A recent example of an ethical clash between researchers and the public is chimera science—"this subset [of scientific research] consists of scientists introducing human cells into animal embryos at a very early stage" ("First human-pig embryos"). In January 2017, Salk Institute researchers in California were successful at creating an embryo that contained both human and pig cells. According to *CNN Wire*, "It is a small but significant step toward the ultimate goal of growing human organs in animals" for transplant purposes. The funding for this project, however, was explicitly denied by the U.S. National Institutes of Health, and the research was subsidized by "private Spanish sources" ("First human-pig embryos").

On the other hand, the public's emotional engagement with science does not always have negative consequences. Modern technology allows for a variety of interactive forums where non-professionals can express their opinions. Quite often, emotional evaluations displayed through interactive commentaries to popular science blogs and online articles are positive, as Luzón (2013) demonstrates. In fact, she shows that scientists welcome the kind of interaction the new media affords: "Scientists ... understand that communication to a public audience is part of developing science and thus seek to explain science, contribute to the public understanding of science, and prompt the public to make decisions based on this understanding" (Luzón 2013: 452). Supper (2014) reports sonification of scientific data (representation of scientific findings with sound) which result in "sublime experiences of science" akin to those that a person experiences after an interaction with a work of art. Fiction literature also responds to the challenge of emotional engagement between the scientific community and the lay public. Scientifically themed novels are on the rise today (see chapter 11).

No matter how one sees the relationship between professional scientists and the lay public, it is clear that the scientific community is in a more powerful position. It has ample choices of popularization channels and through careful use of language can create the exact image of itself it wants the public to have. For example, here is Michio Kaku's (2011: 146) optimistic message about the powers of modern genetics. Kaku is writing about advancements in research on anti-aging: "Scientists have now isolated a number of genes (age-1, age-2, daf-2) that control and regulate the aging process in lower organisms, but these genes have counterparts in humans as well. In fact, one scientist remarked that changing the life span of yeast cells was almost like flicking on a light switch." Kaku is deliberately drawing a comparison between the ease with which one can turn on the light and the process of extending the human life span. With one phrase, "these genes have counterparts in humans as well," he is giving hope that is perhaps not warranted just yet, taking into account actual scientific findings. However, Kaku's point is to present research results in such a way that they make scientists into heroes who are a few experiments away from delivering immortality on demand.

The notion that popular science presents scientists and their work in an exclusively positive light is not new. Several linguistic studies of popularizations make similar claims covertly or overtly: For instance, Fu and Hyland (2014: 123) argue that popular science is "persuasive, seeking to convince the reader both of the importance of the content and a wider ideology of scientific progress." Bucchi (1998) makes a less veiled claim that popular science employs what he labels "'celebratory' discourse"—essentially an approach that celebrates science, its achievements and practitioners rather than offering any kind of critical or analytical perspective. Harré (1994) points out that professional science and the publications it produces are not much different—focusing on positive outcomes of laboratory procedures at the expense of the truth. He further explains, "If anyone tried to publish a story more like real life, in which hypotheses were dropped for lack of support, apparatus couldn't be made to work within the parameters of the original experiment, and so on, it would be turned down" (Harré 1994: 87). "Science," Harré concludes, "must present a smiling face both to itself and to the world" (1994: 87). And so it does. In fact, science has become so good at it that its professional and popular outlets have developed specific linguistic formulae for constructing "a smiling face."

This book, in discussing linguistic features of popular science, will address two specific features that contribute to celebration of science and scientists the most—narratives of discovery and presented discourse. The first are stories that explain how scientific breakthroughs came about; the second are voices and thoughts (most often of scientists) included into texts as direct quotes or paraphrases. The book will not, however, focus entirely on the celebratory features but rather take a broader approach, giving you a guided tour through popular science books and explaining what makes them successful and how they manage to appeal to a broad and diverse audience.

First of all, I must say that this exploration is limited to a discussion of popular science in book form for a reason. Popular science books target an audience that is slightly more educated and sophisticated than that of popular magazines and online articles (Hyland 2010: 118). As such, books allow for a deeper and more detailed exploration of the material. Another reason to look at books as opposed to magazines, newspapers, or online publications is the fact that, as Turney (2007: 2) puts it, "books ... suit science." He means that "scientific explanations are very often highly embedded—one thing depends on understanding several others and there is often a whole web of concepts and entities which have to be introduced to tie the explanation together. And books lend themselves to extended—often *very* extended—many-layered, explanations" (Turney 2007: 2).

In fact, explanation plays a key role in popular science no matter the form, but Turney's assessment is entirely accurate—in popular science books explanation is the key element of almost everything: from organization to the word choice. As I will show, the need to explain, and to explain in a relatable and engaging manner, drives the authors to use the three linguistic features I make the center points of this book. The connection between definitions and explanation is, perhaps, the most obvious.

Authors of popular science utilize a range of ways to present scientific terminology to their readers. Both prototypical definitions (along the lines of A = B) and non-traditional definitions (definitions that show how something works, for example) could be found in a variety of popular science books. Some definitions reflect not only the already established scientific knowledge, but also present the reader with the process of knowledge construction and scientific discovery. According to Chakrabarti (1995:10), "The purpose of a definition is to increase understanding," and the authors of popular science books demonstrate full awareness of such a purpose and come up with creative ways to incorporate definitions into a text without breaking the natural flow of the prose. They may also use definitions to introduce their own points of view on the concepts and objects being defined. Here is an example of a non-traditional definition from Brian Greene's *The Hidden Reality*: "The big bang model describing a cosmos that began enormously compressed and has been expanding ever since became widely heralded as the scientific story of creation" (Greene 2011: 20). This definition contains all the necessary components of a traditional definition. It has the subject (what is being defined—"the big bang model") and the describers (the words that give information about the concept being defined—"a cosmos that began enormously compressed and has been expanding ever since" and "the scientific story of creation"). What makes this definition unusual, and what makes it blend into the surrounding text, is the way its two component parts are joined, or rather how they are not. In a traditional definition, there would be a verb connecting the two parts: something similar to "is" or "is defined as." In the definition above, however, there is no such verb. This definition is less concerned with the prim structure and more with explaining the concept of the big bang. In one relatively short sentence it introduces the scientific explanation of this cosmological model and signals its place among the existing scientific views.

Because this definition is structured less traditionally, it also introduces a degree of doubt about the permanency of the big bang model's status as "the scientific story of creation." The choice of the verb—"became"—implies that things can change. Further addition of "widely heralded" again introduces a possibility that the situation is not permanent—we all know that the acceptance of a majority does not always imply truthfulness. Greene made a conscious decision to hedge this definition. Had he used "is" instead of "became widely heralded," it would have been a much stronger endorsement. In the rest of the book, the reader finds out the reasons behind Greene's hesitation.

Narratives, especially the stories of discoveries, are also guided by the desire to explain. They, however, have many more tools than the definitions at their disposal to make the point and to illustrate it thoroughly. For example, many of the narratives are two-fold: they consist of the basic story line and a secondary, scaffolding structure that supplies the explanation and the evaluation of the events by the author. What's peculiar about this double makeup is that the explanation is in many cases not contained in one neat section but dispersed throughout the story, appearing at the points deemed crucial by the author. Below, I introduce a narrative of discovery from Marcus du Sautoy's (2011: 70–74) *The Number Mysteries*. The narrative has been subdivided into component categories according to Labov's (1972) model of narrative organization (I discuss his framework in more detail in chapter 2). The names of the categories should be self-explanatory:

"Bubbles Fusing Together"

1. Abstract

The first proof that the fused bubbles couldn't be bettered was announced in 1995. Although mathematicians don't really like asking for help from a computer (because that doesn't appeal to their sense of elegance and beauty), they needed one to check through the extensive numerical calculations that were involved in their proof. Five years later, a pencil-and-paper proof of the double-bubble conjecture was announced. It actually proved a more general conjecture:

2.a. Explanation

If the bubbles do not enclose the same volume, but rather one is smaller than the other, then the bubbles fuse together so that the wall between the bubbles is no longer flat but bent into the small bubble. The wall is part of a third sphere and meets the two spherical bubbles in such a way that the three soap films have angles of 120 degrees between them.

3. Orientation

In fact, this 120-degree property turns out to be a general rule for the way soap bubbles fuse together. It was first discovered by Belgian scientist Joseph Plateau, who was born in 1801.

4.a. Complication

While he was doing research into the effect of light on the eye, he stared at the sun for half a minute, and by the age of 40, he was blind. Then, with the help of relatives and colleagues, he switched to investigating the shape of bubbles. Plateau began by dipping wire frames into bubble mixture and examining the different shapes that appeared.

2.b. Explanation

For example, when you dip a wire frame in the shape of a cube into the mixture, you get 13 walls that meet at a square in the middle. This "square," however, isn't quite a square—the edges bulge out.

4.b. Complication

As Plateau explored the various shapes that appeared in different wire frames, he began to formulate a set of rules for how bubbles join together.

5.a. Result

The first rule was that soap films always meet in threes at an angle of 120 degrees. The edge formed by these three walls is called a Plateau border in his honor. The second rule was about the way these borders can meet.

2.c. Explanation
Plateau borders meet in fours at an angle of about 109.47 degrees ($\cos^{-1}-1/3$, to be precise). If you take a tetrahedron and draw lines from the four vertices to the center, you get the configuration of the four Plateau borders in foam. So the edges in the bulging square at the center of the cube wire frame actually meet at 109.47 degrees.

5.b. Result
Any bubble that didn't satisfy Plateau's rules was believed to be unstable and would therefore collapse to a stable configuration that did satisfy these rules.
It was not until 1976 that Jean Taylor finally proved that the shape of bubbles in foam had to satisfy the rules laid down by Plateau.

6. Coda
Their work tells us how the bubbles connect together, but what about the actual shapes of the bubbles in foam? Because bubbles are lazy, the way to the answer is to find the shapes that enclose a given amount of air in each bubble in the foam while minimizing the surface area of soap film.

Explanations permeate the narrative and often interrupt the flow of events in order to interject this or that piece of information or to clarify what has been said. The fact that an explanatory element has become a structural feature of discovery narratives underlines its importance (see pages 63–64 for more details).

Voices of scientists are also frequently used to generate explanations. Most often, it is indirect discourse—speech that has been summarized or in other ways reformulated by the author—that delivers explanations. Here is an example from Bill Bryson (2003: 139–140). He is writing about Rutherford's discovery of a structure of an atom: "An atom, Rutherford realized, was mostly empty space, with a very dense nucleus at the center." When it comes to explaining the experimental procedure, Bryson (2003: 139), instead of supplying his own account, chooses to let Rutherford himself describe it for the readers: "It was as if, he [Rutherford] said, he had fired a fifteen-inch shell at a sheet of paper and it rebounded into his lap" (Bryson 2003: 139–140). The author uses Indirect Speech of the scientist to introduce the explanation. Having read this sentence, a reader would have no doubt that it was Rutherford that came up with the analogy; however, evidence suggests that this particular explanation is a rhetorical device very likely invented by popularizers and not by Rutherford. (More on this and other aspects of presented discourse in chapter 4.)

Carefully crafted explanations of complex scientific ideas, on the other hand, do not alone account for the success of popularized material. Infusing science with humor and personal anecdotes never fails. You might be surprised by the number of humorous asides and straight out funny stories that make up the world of popular science. What is more surprising is that quite often the authors deliberately choose to relate serious and not necessarily funny material in a light-hearted way. For example, here is how Greene (2011: 38) describes George Gamow's defecting from the Soviet Union (not a laughing matter at all):

> (in 1932, he and his wife tried to defect from the Soviet Union by paddling across the Black Sea in a kayak stocked with a healthy assortment of chocolate and brandy; when bad weather sent the two scurrying back to shore, Gamow was able to fast-talk the authorities with a tale of the unfortunately failed scientific experiments he'd been undertaking at sea). In the 1940s, after having successfully slipped past the iron curtain (on dry land, with less chocolate) and settled in at Washington University in St. Louis, Gamow turned his attention to cosmology.

Biographical details of scientists aside, hard scientific facts do not escape humorous treatment either. Kean (2012: 151) uses the following description to illustrate viruses' ability to alter a host's DNA: "a lab-tweaked version of one virus can turn polygamous male voles—rodents who normally have, as one scientist put it, a 'country song ... love 'em and leave 'em' attitude towards vole women—into utterly faithful stay-at-home husbands, simply by injecting some repetitive DNA 'stutters' into a gene that adjusts brain chemistry." Humor, of course, is meant to appeal to the lay reader and to put her at ease with what might otherwise seem a complicated subject matter—it is a type of interaction with the reader (see chapters 9 and 10 for more on the relationships between readers and writers).

Another way to make science appealing and, at the same time, to demonstrate that it is not entirely separate from the world of the arts is to include literary references as parts of explanations. This is much more likely to happen in popular science books than in other media (print and internet articles, blogs). Length of a popularization might be a factor; however, the audience is a more likely influence. Readership of popular science books has been reported (see, for example, Hyland 2010) to include people with higher levels of education as compared to those who get their popular science from blogs and magazines. As such, this kind of reader is more likely to appreciate mentions of literary classics. Take this example: "As well as responding to individual events, animal nervous systems can also respond to sequences of events over time. If you repeatedly stimulate the slug's siphon, the gill-withdrawal reflex progressively weakens.... It is a case of what Marcel Proust called *the anaesthetizing effect of habit*" (Coen 2012: 141). The references to Proust, Shakespeare, Tolstoy, and the like are also mechanisms for the authors to tell the reader something of themselves—to showcase their own knowledge of an area that lies outside their immediate expertise. In fact, personal stories, especially those that recall how one became interested in science, are a staple of popular science. Like humor and literary references they are designed to humanize science. In the end, this book will demonstrate that all of the linguistics techniques that make up successful popular science writing are in some way connected to convincing the reader that the people wearing lab coats and goggles or the ones staring at the night sky through telescopes are not all that different from the one who is reading about them.

1

A Linguist Looks at Popular Science

You may be wondering why popular science would interest a linguist. Well, for this linguist, it was a personal matter. Growing up a child of chemists, I was expected to choose a career path in the sciences. Physics, math, chemistry, biology, etc., were all acceptable options to my parents. They were in for a surprise when in the ninth grade I chose to specialize in foreign languages and literature—an opportunity my school offered alongside specializations in physics and math or biology and chemistry. Upon conveying my decision to my parents, I was promptly told that all the smart kids (no matter their real interests) would go into the physics and math group and that I should join them lest I ended up among the academic outcasts and underachievers. I stood my ground and thus began my long, winding path to a PhD in applied linguistics, which my parents have since come to see as a respectable career choice.

In a way, looking at the language of popular science is a personal tribute to my parents and to the many scientists whose ranks I chose not to join but whose work I admired throughout my life. Examining the language of science written for non-specialists also answers some important questions: Why are these books so well received? Why do we love them when we hated our science textbooks? This type of research may also help scientists deliver their exciting (and sometimes not so exciting but very valuable) knowledge to the public.

I devote this chapter to an exploration of a limited number of linguistic features commonly found in popular science. First of all, I will discuss the role storytelling plays in popular and professional science and in the area of information transfer in general. Second of all, I will examine a framework that allows popular science authors to include voices of scientists into their books. Third, I will talk about the use of scientific terminology in a text designed for non-specialists. All these aspects will receive full attention in the chapters that follow. Here, I mean to introduce some theoretical background on each one of them and to prepare you for more in-depth discussions.

Let's begin with narrative. Everyone's familiar with stories. We hear and tell stories everyday. There is, it appears, nothing special about them. There is and there isn't. The reason we feel so comfortable with narrative is because we are wired to process information that way. Research in neuroscience, evolutionary psychology, artificial intelligence, narratology, and linguistics all points to the idea that human brains are predisposed to information that comes in structured as a story.

Stephen Hawking had to learn this the hard way. While working on his now famous book *A Brief History of Time*, he shared a draft of a section with Simon Mitton, a fellow scientist and a writer. When he read the draft, Mitton uttered probably the most famous piece of advice to a popular science writer: "It's still far too technical, Stephen.... Look at it this way, Steve—every equation will halve your sales." Hawking wasn't convinced, so Mitton had to explain, "Well ... when people look at a book in a shop, they just flick through it to decide if they want to read it. You've got equations on practically every page. When they look at this, they'll say, 'This book's got sums in it,' and put it back on the shelf" (cited in White and Gribbin 1992: 222–223). Hawking followed his colleague's suggestion and ended up with a super best-seller. According to Hawking's foreword to the 1996 edition of the book, "It was in the *London Sunday Times* best-seller list for 237 weeks, longer than any other book (apparently, the Bible and Shakespeare aren't counted). It has been translated into something like forty languages and has sold about one copy for every 750 men, women, and children in the world."

So what is the secret? Why do we crave stories and resent equations? There has to be something more to it than memories of math classes. And after all, not all of us felt great joy and exaltation doing homework for English classes either. Psychologists argue that even our mental well-being is connected to our ability to tell stories. According to Baerger and McAdams (1999), those of us capable of constructing a coherent life story out of personal experiences show greater signs of psychological well-being than those who cannot tell a well-put-together tale. According to researchers in such varying disciplines as artificial intelligence and literary criticism, narrative structure is innate for humans.

Roger Schank (1990: 40–41), someone who has been pondering the questions of artificial and real memory for a number of years as a scientist, a psychologist, and an educator, has determined that we are so fond of stories because, as he put it, "stories are usually told to someone and not to an empty room" and allow us to showcase our knowledge and our selves to fellow human beings. They serve our communicative goals—and we pretty much live to communicate with our fellow humans. In Schank's (1990: 29) words, "Every human communication revolves around stories." And stories are, as we will see a bit later, highly subjective.

This communicative approach to information is not accidental. There is a whole branch of linguists called Integrationalism that approaches language not as abstract entity but as a set of concrete occurrences specific to certain communicative situations and the people involved in them. This movement began with Roy Harris, an Oxford professor of linguistics, and now continues through the International Association for the Integrational Study of Language and Communication and most notably through Professor Michael Toolan, author of *Total Speech: An Integrational Linguistic Approach to Language*. One of the best and clearest examples of linguistic subjectivity, as Toolan (1996: 8) notes, is the law. In his own words, "We may imagine that we have a pretty clear idea of the literal or conventional meaning of such words ... as *chicken*, *fire*, *souvenir*, *race*, and *sausage*. But various commentaries have charted in some detail how minor and major judicial cases are in practice determined in highly contextualized ways.... In all these cases, the criteria for what counts as a souvenir or a sausage emerge from the particular circumstances of each controversy." Such a view of language suggests that we have a predisposition not for the objective and abstract but for the subjective and contextual. It comes as no surprise then that we prefer stories to dry facts since narratives are excellent at supplying information in context and adding personal interpretations.

As George Zarkadakis (2016: 18), a noted expert in artificial intelligence puts it, "Language evolved ... as a means of enhancing social cohesion." He adds, "The main purpose of language was, and still is, gossip." That is telling stories about ourselves and others. This might seem unremarkable. After all, the communicative nature of language is obvious. However, the idea of developing a language for the purpose of storytelling is worth commenting on. We humans are not telling stories because we have the linguistic ability to do so. As Brian Boyd (2008) points out, "storytelling exists as [a] cross-species" behavior.

From a strictly linguistic point of view, however, only humans are capable of fully fledged narratives. A famous example often given to illustrate other species' narrative prowess is a honeybee's dance which involves tail-wagging to indicate that it has found a source of nectar. Such a communicative act, in Toolan's (2001: 5–7) words exhibits one characteristic of narratives—spatial displacement (the ability to talk about things or events that are not present or happening right now)—but it lacks the ability to relate the same information at a later time—temporal displacement. A honeybee can only communicate with its hive-mates immediately after it has located a source of nectar. It has to tell its "story" right away; it can't wait until the next day or until the right moment comes.

The reason Boyd (2008) is able to make the statement about the interspecies use of narrative is because he regards storytelling in broader terms—as a manifestation of art. Art is all about patterns and play, he argues, and humans are by far not the only occupants of this planet to engage in pattern recognition, play, and therefore art, which in turn leads to communication through stories. In his book *On the Origin of Stories*, Boyd (2009: 129) makes a point that "narrative depends heavily on understanding events, and event sequences"—that is what the honeybee is capable of doing. As Boyd (2009: 129) continues, "We share much of that understanding with other species, but it takes important new forms in humans." Unlike the honeybee, we have evolved not only to understand the world around us well enough to be able to express it as an event sequence, but we are also capable of producing and recognizing representations. And, in fact, we prefer fictional stories to narratives that reflect reality verbatim.

Some researchers (me included), however, have argued that fiction permeates all stories no matter their genre or communicative intent—as narratologists Skov Nielsen, Walsh, and Phelan put it, fictionality is "ubiquitous" (2015a: 62) and "extremely pervasive" (2015b: 110). At the same time it remains a phenomenon associated primarily with fiction. Skov Nielsen and his colleagues assert that "apart from the work by literary critics on generic fiction, fictionality is almost completely unstudied and often unacknowledged" (Skov Nielsen et al. 2015a: 62). At the same time, attempts by artificial intelligence experts to replicate the human mind have lead the researchers to accept the pervasiveness of fictionality and the important role it plays in information delivery and processing. Schank (1990: 44) argues that stories by their nature are fiction. "Our stories," he declares, "because they are shaped by memory processes that do not always have their basis in hard fact, are all fictions." Schank (1990: 44) hastens to add that "these fictions are based on real experiences and are our only avenue to those experiences." Zarkadakis (2016: 24) offers further explanation: "Our memories are not precise recording instruments. Our brain is not like the hard drive of a video camera. Every time we describe a past event, our brain recalls a few facts and automatically fills the gaps with whatever can be used to preserve the coherence of our narrative."

In other words, even the most credible non-fiction text contains elements of fiction. Linguistics can show exactly how and where this happens. One favorite location of fictionality, as I came to discover during my PhD research, is quotes and paraphrases that present speech and thoughts of people discussed in a narrative. The technical term for such passages of text is *presented discourse*.

Storytellers, to no one's great surprise, take liberties with presenting the speech of others. As Toolan (2001: 128) points out, such fictionalization is most likely to happen in what is known as *indirect discourse*—presentation of speech that is not a direct quote but rather a paraphrase; it is most often preceded by a word "that," as in *Mike told me that he will not make it in time for dinner*. Now, Mike did not have to use the words "make it" or "dinner" for my presentation of his utterance to convey the same meaning. He might have said, *I will be unable to come home in time for the evening's meal*. This is a very minor adjustment, however. It is entirely possible that Mike never said anything even remotely like my paraphrase. He could have said, *Don't wait up; I'll be late*. Knowing the context, I construed the paraphrase that fit better into my narrative than what the speaker actually said. Toolan (2001: 128) takes this notion further, saying that "people are quite capable of 'reporting' things that their reportees never said." Deborah Tannen (2007) talks of whole reported dialogues that never happened. She uses the term *constructed dialogue*. Remember, we are not talking about novels or plays here; this happens in everyday conversational storytelling and in non-fiction texts, like popular science books.

What is going on? Are we compulsive liars? No. "Just natural storytellers," according to Zarkadakis (2016: 24). In 2011, Michael Gazzaniga, a neuroscientist, reported an experiment he and his colleagues had conducted that confirmed "the neurological basis of storytelling" (Zarkadakis 2016: 23). The experiment involved patients whose right and left hemisphere of the brain were split and did not communicate. Unable to detect that certain activities were initiated at the request of one hemisphere, the other hemisphere, Gazzaniga showed, had to invent a reason why a certain bodily movement occurred—in effect, producing a story.

The linguistic explanation for constructed dialogue or presentation of speech dissimilar to the actual utterance is offered by Tannen (2007: 120): "The speaker uses the animation of voices to make his story into drama and involve his listeners." Dramatization and fictionality, in other words, go hand in hand. This is a connection that I will explore in chapter 5, which is devoted to presentation of speech. For now, let's continue with the more general discussion of narrative.

As natural storytellers, we let narratives permeate our lives. Researchers noticed this in the late 1980s and early 1990s (e.g., Kreiswirth 1992). The fascination with stories as a means of interdisciplinary inquiry has been dubbed *narrative turn* (e.g., Hyvärinen 2010). There is a myriad of studies that explore the application of narrative to such different fields as accounting (e.g., Sydserff and Weetman 1999) and healthcare (e.g., Balfe 2007, Gülich 2003), for example. However intriguing such explorations are, for the purposes of this book it would be more productive to address the use of narrative in professional sciences before moving on to a discussion of storytelling in popularizations.

When I look at the titles of science communication books (the kinds of publications that explain to professionals how to get their message across to a non-specialist audience) that came out in the last couple of decades, it is very tempting to say that professional sciences have experienced a narrative turn—just like the humanities and the social sciences. It is, however, grossly untrue. First of all, there has not always been such a thing as Professional Science—it is a deliberate social construct of the 19th century (this particular discussion started in the introduction and continues in the conclusion). Second of all, modern professional science has always maintained close ties with narrative conventions.

As Harré (1994) points out in his analysis of professional scientific publications, research articles follow narrative and character-creating conventions similar to those one might come across in a short story or a novel. For instance, Harré (1994: 86) suggests that it is entirely legitimate to designate the scientist (or a group of researchers) who performs the experiment described in the article as a "hero," using the literary sense of that term. He goes on to say that this hero follows three predictable stages: first, he "presents a hypothesis," then he tests the hypothesis and obtains empirical support, and finally the hero presents the results of his experiments as proof for the initial hypothesis. This is a neatly ordered, chronological sequence of events, what narratologists and linguists call a *temporal sequence*—the main minimum requirement for a traditional a narrative.

Randy Olson (2015) in his book *Houston, We Have a Narrative: Why Science Needs Story*, boils down the temporal sequence of a typical research article to a memorable acronym: IMRaD, which, according to his own account, very few scientists are able to recognize until it is expanded into the Introduction, Methodology, Research, and Discussion. Is a story structured according to these guidelines different from the one in the pages of a novel? Absolutely, but that does not make it a non-story. So while the surface elements of professional scientific narratives (such as words and grammar, for example) are different from literary fictions, the underlying narrative structures are quite similar. Popular science, as I will demonstrate, is even closer to fiction in a number of structural and surface elements.

The emotional impact that a good story produces is also a unifying link between professional and popular scientific narratives and literary fictions. When we communicate, be it our findings about the cosmic microwave background radiation or the quality of room service in a nearby Hilton, we want to tell a story that is memorable and has a positive outcome—a resolution, a conclusion that leads somewhere and introduces a solution. In professional scientific publications this chase for the positive outcome of research, some argue, has begun to damage the integrity of the research process. In other words, scientific findings are much more likely to be published if they introduce a positive outcome to the experiment described. As Harré (1994: 87) writes, "If anyone tried to publish a story more like real life, in which hypotheses were dropped for lack of support, apparatus couldn't be made to work within the parameters of the original experiment, and so on, it would be turned down." Olson (2015: 9) attributes this trend to the basic need of attracting audiences through good storytelling, which he equates with positive outcomes of scientific research: "The positive result is the same as telling a good story." Communication of science, Olson (2015: 10) demonstrates, favors the impact a story will have over the "soundness" of the research that went into a particular scientific result.

For professional science, however, close contact with stories does not have to be negative. The general idea of the presentation of research findings in narrative form is invaluable and has been proven effective. In fact, quite a few researchers believe that the real power of stories lies not necessarily in our innate ability to produce and process information in narrative form but in a narrative being a superb platform for context necessary to interpret facts (e.g., Avraamidou and Osborne 2008, Reitsma 2010, Turney 2004b). As Reitsma (2005: 93), writing about geosciences, notes, "A narrative ... includes interpreted information"; he argues that this eliminates the need to draw conclusions among the events presented.

Of course, science in narrative form is also vital for the popularizing outlets of modern research institutions, as I have already mentioned in the introduction. Narratives allow for a development and an exploration of personal connections between research objects and goals and the public. Stories help the public explore novel and sometimes alien (or alienating) issues in familiar contexts. For the scientists, they create an opportunity to look at research from a new perspective. Moreover, presenting science in the form of a story lets the public see scientists as relatable people. I will touch more on that relationship in chapters 2 and 3, as well as in the conclusion. For now, let's move on to a brief introduction to presented discourse.

I mentioned presented discourse above, saying that it amounts to quoting what someone said or paraphrasing the speech of others. There is more to the idea of presented discourse, however. First of all, it doesn't deal with speech only. A writer or speaker can easily present the thoughts of others. Second of all, more recently, linguists have started talking about the presentation of writing alongside presentation of speech and thought. Elena Semino and Mick Short (2004) were instrumental in establishing presentation of writing as a viable category of presented discourse (see chapter 7 for more).

When discussing presented discourse, it is essential to be aware not only of the three major categories (presentation of speech, presentation of thought, and presentation of writing) but also of the forms that presentations of various voices in a text may take. A major study of the manifestations of presented discourse was conducted by Geoffrey Leech and Mick Short in 1981 with a revised edition coming out in 2007. It was called *Style in Fiction*. As the title suggests, the authors examined how speech and thought (but not writing) were represented in novels and short stories.

The value of Leech and Short's model is that it presents clear analytical categories for both speech and thought. As Short (2007: 226) observes, "It was the first attempt to distinguish systematically between speech and thought presentation." In the overview that I introduce below, I use primarily the Leech and Short model; however, I do include explanations that reference a later study by Semino and Short (2004). I discuss the categories of speech and thought presentation together to highlight the fact that each category of speech has a counterpart in the presentation of thought and vice versa. There are five categories of speech presentation and five corresponding categories of thought presentation in Leech and Short's (1981/2007) model. In addition to full category names, there also exist agreed-upon abbreviations for each. I include them in parentheses. When I discuss presented discourse in detail (chapters 4, 5, and 6), I will use these abbreviations but will also include the full category names from time to time.

When speech or thoughts of others enter into a text, the narrator usually uses a mechanism to help the readers or listeners distinguish those words from the main story. The phrases that help accomplish this are called reporting clauses. Here are two examples: *she whispered* or *he thought*. Both of these phrases signal that the words that follow belong to someone other than the narrator. In the examples below (that come from the popular science books analyzed in the later chapters), the reporting clauses will be underlined.

Now let's look at the most commonly recognized ways to include speech or thoughts in a text:

> **Direct Speech/Thought (DS/DT)**—The original utterance or thought presented as if it were verbatim and introduced by a reporting clause. *Effects produced*: Focus on the original speech or thought situation and faithfulness of representation. Emphasis on the originator of the speech or thought [Leech and Short 2007: 256–257].
>
> **Example**: <u>Rous himself later admitted</u>, "I used to quake in the night for fear that I had made an error" [Kean 2012: 141].

An example of Direct Thought would have a mental process verb (a verb that expresses an act of thinking) rather than a speech verb: *realized* instead of *admitted*, for instance.

There aren't very many verbs that are used to express mental processes in reporting clauses. The verbs that are used most often are *to wonder, to come up* (with an idea), *to think, to realize,* and *to assume*. Both Direct Speech (DS) and Direct Thought (DT) can be introduced without a reporting clause. In that case, the instance of presented discourse is called Free Direct Speech or Free Direct Thought:

> **Free Direct Speech/Thought (FDS/FDT)**—Original discourse presented as if it were verbatim but without the reporting clause and often without the quotation marks. *Effects produced*: Focus on the character's voice without "the narrator as an intermediary" [Leech and Short 2007: 258].

Some researchers see no significant difference between Free Direct Speech or Free Direct Thought and their Direct counterparts, so they suggest combining the two categories. For example, according to Semino and Short (2004), the presence or absence of quotation marks and/or reporting clauses does not affect how the reader perceives the stretch of discourse presented. They argue that most instances of dialogue omit the formal marks of direct discourse (that is quotation marks), yet do not aim at a different effect by doing so. The acronyms for the combined categories are (F)DS and (F)DT. Most of (Free)Direct Speech in the popular science books I analyzed includes quotation marks but does not use reporting clauses—thus representing a mixture of Direct Speech and Free Direct Speech—which tends to argue for Semino and Short's (2004) combined category. You can try to spot Free Direct Thought in the following example:

> The healthy plants, <u>Mayer discovered</u>, turned sick as well. Some microscopic pathogen must be multiplying inside the plants. Mayer took sap from sick plants and incubated it in his laboratory [Zimmer 201: 3–4].

The second sentence is a classic example. It does not have a reporting clause, nor are there quotation marks, yet we can tell that it expresses Mayer's thoughts and not the ideas of the narrator. The hint is the use of the present tense—*must be* instead of *was*—it creates a sense of immediacy common to Direct Thought, yet there are no quotation marks or reporting clause to mark the Direct Thought; therefore, it can be labeled Free.

At this point, we have reached a very important dividing line among the categories of presented discourse. It is a line between Direct and Indirect discourse. The division is so significant that some researchers, for some forms of linguistic analysis, choose to focus on just the categories of Direct or Indirect discourse, disregarding the finer distinctions. For instance, Waugh (1995) analyzes only Direct and Indirect Speech in newspaper reports. Myers (1999) chooses to emphasize direct discourse when examining presented speech in oral group discussions. For Urbanova (2012) it is sufficient to examine only free direct and direct forms of presented discourse, which also suggests an underlying broad contrast between direct and indirect discourse.

Up till now, we have been looking at the forms of Direct Discourse: Direct Speech, Direct Thought and their Free variants, Free Direct Speech and Free Direct Thought. Indirect discourse adds three more categories:

> **Indirect Speech/Thought (IS/IT)**—Reformulation of an original utterance or thought that contains a reporting clause. *Effects produced*: More complete integration into a narrative compared with DS/DT; focus on the message rather than on the exact words [Leech and Short 2007: 256–257].

You will recognize Indirect Speech or Indirect Thought as paraphrases.

> **Example**: Two thousand years later, <u>the physiologist Leonard Hill argued</u> in the 1920s that they [colds] were caused by walking outside in the morning, from warm to cold air [Zimmer 2011: 10].

The underlined reporting clause indicates that a different voice from the narrator's has entered the text; at the same time, the absence of quotation marks suggests that it is just the general idea of Leonard Hill that is being introduced and not exactly his own words. This is a prototypical example of Indirect Speech, where the reporting clause precedes the reported utterance. In my analysis, however, I have come across some utterances (but not thoughts) that are interrupted by a reporting clause. For example:

If something like water is heated, so that it evaporates and turns into a gas, the same corpuscles would still be there, said Boyle, but the gas occupies more space than the liquid had done [Bynum 2012: 85].

The next category of indirect discourse is Free Indirect Speech and Free Indirect Thought. These categories are considered sufficiently different in the effects they produce from Indirect Speech and Indirect Thought and therefore do not combine with them.

> **Free Indirect Speech/Thought (FIS/FIT)**—Indirect Speech or Indirect Thought that is presented without a reporting clause. Free discourse reflects the narrative's perspective from the point of view of the character, which sometimes results in tense shifts. *Effects produced*: While not a faithful reproduction of the original utterance or thought, it still has more power to refer to the feel of the original than Indirect Speech or Indirect Thought [Leech and Short 2007: 261].

Free Indirect Speech may produce confusion, as it obscures the identity of the speaker and requires the reader to pay extra attention to determine who is talking. Here is an example of FIS from a popular science book (Free Indirect Speech is in bold):

> Far from rejoicing, the older scientist screwed up his brow and expressed his doubts that the nucleus contained any sort of special, non-proteinaceous substance. **Miescher had made a mistake, surely**. Miescher protested, but Hoppe-Seyler insisted on repeating the young man's experiments—**step by step, bandage by bandage**—before allowing him to publish [Kean 2012: 20–21].

The tense shift from the simple past to the past perfect indicates a change in the narrative focus and introduces some degree of immediacy and surprise that the reader is to associate with Hoppe-Seyler. In this example, the identity of the speaker is less obscured than in some; however, it is still possible to confuse this instance of Free Indirect Speech with narration—the voice of the narrator.

The next category will receive much attention in the later chapters. Of all indirect forms of presented discourse, it is the most author-controlled, that is, it presents the most generalized idea of an utterance or thought. Leech and Short (1981/2007) call this category of presented discourse Narrative Report of Speech Acts/Narrative Report of Thought Acts. Semino and Short (2004) label it Narrator's Representation of Speech/Thought Acts, and Short (2012), on the other hand, makes a case for using the term "Presentation" instead of "Representation" and the abbreviation "P" instead of "R." I will not go into the details of this substitution of terms, yet I will use Short's (2012) name for the category—Narrator's Presentation of Speech Acts/Narrator's Presentation of Thought Acts.

> **Narrator's Presentation of Speech/Thought Acts (NPSA/NPTA)**—Summaries of utterances or thoughts. *Effects produced*: Deemphasizing of the importance of the

original utterance or thought in the new context. Emphasis on the fact that a speech or thought act took place neither on the words nor on the message [Leech and Short 2007: 259–260].

These are all of the presented discourse categories in the Leech and Short (1981/2007) model. Semino and Short (2004) introduce several more which are, in many cases, designed to reflect very subtle distinctions. For the analysis of presented discourse that appears in chapters 5–7 I will use only the categories listed above. However, if you would like a more in-depth look at presented discourse you can consult Semino and Short (2004) or Pilkington (2018).

Leech and Short (1981) are also known for the introduction of the speech and thought presentation scales that arrange the categories of discourse presentation according to the degree of authorial control from the most controlled to the least controlled. Author control, in this case, means how much is changed in the original utterance or thought between when it was produced and recorded. An instance of Direct Speech, for example, has the least or no change—the words are reported verbatim (with some possible omissions indicated by an ellipsis). Narrator's Presentation of Speech Acts, on the other hand, requires a lot of change because no exact words are being used; instead the author supplies a summary.

The combined scale for speech and thought is shown below with the Narrator's Presentation of Speech/Thought Acts (NPSA/NPTA) being the categories most controlled by the author and (Free)Direct Speech/(Free)Direct Thought being the least controlled:

NPSA/NPTA—IS/IT—FIS/FIT—(F)DS/(F)DT

Leech and Short (1981/2007: 276) use this scale to demonstrate that the modes of speech/thought presentation form a continuum, with each category responsible for different effects on the reader depending on the involvement of the author. Leech and Short also mark what they call "the norm" for speech and thought presentation. Keep in mind that they considered examples from fiction only. Thus, the norm for speech presentation in fiction is Direct Speech, and the norm for thought presentation is Indirect Thought. Leech and Short (1981/2007: 276) explain that these categories are chosen as the norms because each of them reflects presented discourse in the form it is directly expressed to the addressee. In other words, speech is directly expressed as Direct Speech, and thought, being an internal process, is accessible to others only via its indirect form. The norms, therefore, reflect the reality of typical interactions.

The last preliminary theoretical section that prepares you for the chapters to come concerns definitions. Here I discuss definitions as they have been approached by logicians, philosophers, and, of course, linguists.

Concise and easy to understand definitions of scientific terminology are probably among the most recognizable features of popular science. At the same time, not every definition that a reader of a popularization will encounter is going to be a short parenthetical aside. In chapters 8 and 9, I will show how complex definitions in popular science can be while at the same time remaining comprehensible. It's all about structure. Definitions in popular science can even be used as interactive segments of the text, where the reader and the author engage in a kind of conversation. Before we move on to this discussion, however, it would be helpful to examine the theoretical underpinning of definition studies, just like we have done with narrative and presented discourse.

There are several academic disciplines that contributed extensively to our understanding of what it means to define something and what shapes definitions take. We will begin with philosophy and logic.

If you ask a philosopher about definitions, he might simply shake his head helplessly since there is a multitude of various approaches to definitions within logic, and most of them contradict each other. One person to attempt a systematic study and a classification of definitions is Richard Robinson—a philosopher and a translator of Aristotle and Plato. His definitive study called simply *Definition* came out in 1950 and then went through several reprints. According to Robinson (1962: 1–11), the problem of how to describe definitions has been debated for "nearly two and a half millenniums"; the debate is not so much about the function of definitions as it is about their application, categorization, and interpretation. In other words, what constitutes a definition is still an undecided issue. Robinson (1962: 2–3) lists 13 various statements of what a definition is. Speaking from the perspective of a philosopher, Robinson (1962: 1) asserts that the multiple theories of definition are often "conflicting" and "bewildering." It is hard to single out one set of ideas that would be acceptable by every theoretician. In fact, Harris and Hutton (2007: vii) describe Robinson's work as "the last general book to be published in English on the theory of definition." This leaves the discussion wide open for the possibilities of new classifications and structures of definitions. Chapter 8 is a result of my own contribution to the definition debate, as I introduce a new model for definitions found in popular science books.

If we look back at the very first attempts to understand what makes a good definition, we will see that they come from the people who were concerned not only with abstract philosophical questions but who tried to understand the world in all of its manifestations. The familiar distinctions between science, philosophy, and logic were not present at the time the scientific method was discovered. The same people who were curious about the arrangement of the natural world were preoccupied with the nature of reality and its linguistic expression. Many of the Ancient Greek philosophers were also the first scientists and scholars of language. For example, Plato (1961 [Hamilton and Cairns eds.]) renders Socrates' observation on the etymology of Greek language in *Cratylus*, and Aristotle is famous for his contributions to mathematics and physics as well as to philosophy and logic.

In fact, Aristotle's view of how the natural world should be analyzed and classified is very similar to his view of how knowledge should be expressed in a definition. In *Physics*, Aristotle suggests that the best way to examine "principles, causes, or elements" of a thing would be to "advance from universals to particulars" (1984a: 315). His view of a proper definition is quite similar. He sees definitions as representing both universal (or general) features and particular (or specific) features of the objects they define. The universal features he labeled *genus* (or class), and the particular features he called *differentia* (or difference). Aristotle argues, therefore, that definitions must include genus, but also comments that difference is easier to identify: "It is easier to define the particular than the universal—that is why one should always cross from the particulars to the universals" (1984b:161–162). In other words, it is easier to identify what makes something different from other objects than it is to determine how this object fits with others. A classic, or prototypical, definition that comes out of Aristotle's observations would have the following structure: A = B, where A represents the subject being defined (known as *definiendum*), and B corresponds to what is called *definiens* (or describers composed of class and difference).

Socrates, like Aristotle, accepted defining by class and difference. He is famous for positing the question "What is X?" as a prompt to a proper definition. However, according to Santas (1999: 101–102), Socrates recognized only certain possible ways to structure a definition. For example, Socrates approved of a definition only if the defined subject was presented in its singular form and if the subject and its describers "appear to be abstract expressions." Santas (1999: 103–104) demonstrates that definitions found in Socrates' dialogues support Socrates' idea that definitions embody knowledge. Socrates states, "When I don't know what a thing is, how can I know its quality?" (Plato 1956 [Warmington and Rouse eds.]: 29).

While this statement exemplifies the importance of definitions, some scholars see it as problematic. For example, Ahbel-Rappe (2009: 68) presents this approach as "the 'Socratic Fallacy.'" Such "priority of definition" makes it impossible to derive any knowledge in some cases (Ahbel-Rappe 2009: 69). For example, if you do not know what gravity is and cannot find a definition of it, how are you to learn about it? A common answer would be by observing examples or instances of gravity you can possibly compile a definition. However, as Ahbel-Rappe (2009: 69) explains, "for Socrates 'it is impossible to search for a definition F by means of examples of things that are F.'" In other words, definitions by example are not appropriate. Santas (1999: 102) elaborates, "Definitions by example are explicitly excluded by Socrates, on the ground that they are not answers to his question [What is X?]. Socrates rejects a list of examples as an answer … by amplifying his primary questions into their long version [What is the same and what is different in all things that are X?]; and to these questions a list of examples would not count as an answer" (Santas 1999: 102). For Socrates, if it is impossible to arrive at a definition by class and describers, then the defining process and the resulting definition are rejected. As you can tell, Socrates was quite rigid when it came to definitional structures.

Socrates' and Aristotle's views continue to influence modern theories of definition and probably underline the idea of the *isolatable meaning*, that is, the ability of a definition to take place of the subject being defined in a particular text. As Moon (2009: 11) puts it, "it should be possible to replace the item … in its context, by the definition without any loss or change of meaning." Here is an example of a prototypical definition created using the class and difference:

> The place-value system … is a way of writing numbers so that the position of each digit corresponds to the power of 10 that the digit is counting [DuSautoy 2011: 20].

Can you spot which part is the class and which is difference? The class is represented by the phrase "a way of writing numbers," which gives you very general information about the subject (the place-value system). The difference—the end of the definition—provides more specifics.

You are likely to come across this kind of definition in a dictionary; other texts, especially popular science, use many different ways to construct definitions. One of them is to define by introducing examples—a type of definition explicitly rejected by Socrates but probably somewhat acceptable to Aristotle. Aristotle, unlike Socrates, suggests that definitions can be connected to demonstrations: "The principles of demonstrations are definitions" (1984b: 149). As Deslauriers (2007: 90) explains, "When we know something of what a thing is, we can know that it is." Thus Aristotle suggests that demonstrations determine if the object of definition has a physical existence or not. To me, this seems to be the first seed of the separation between definitions of things and definitions of words—the separation essential to some modern theories of definition.

Aristotle cautions in *Posterior Analytics* that "neither is there demonstration of everything of which there is a definition, nor is there definition of everything of which there is demonstration" (1984b: 149–150). This is one of the central points in Aristotle's approach to definitions—human knowledge and understanding are limited, and not everything observable is known or, at least, not known yet.

Nowadays, definitions by example are accepted and called *ostensive definitions* (Robinson 1962: 15, 119–120). Pointing to an object or giving a set of examples has become a legitimate way to define. Such defining process corresponds to Aristotle's idea that defining by difference (that is outlining specific features) is much easier than providing a general description followed by a list of specifics. If you were to ask me, "What is a dog?" and we happen to be at a dog park, it is a lot easier for me simply to point to one specific canine and say, "That! That's a dog." It would take a bit longer to come up with "A dog is a mammal who is carnivorous, has been domesticated, has four legs, is covered with fur," etc.

An ostensive definition (a definition by example or pointing) offers the particular (one specific manifestation) and leaves the listener/reader to arrive at the whole on his own. Some definitions in popular science books can be described as related to ostensive definitions since they define by stressing very specific features.

To sum up, ancient philosophers gave us the basic definition, which has endured to this day in a form of dictionary definition and became the stepping stone for a variety of defining approaches. Now let's look at what the study of logic has to offer to definition practices.

To ancient philosophers, definitions were crucial to epistemology. Knowing a definition enabled one not just to demonstrate knowledge but also to argue about truth and falsehood. Out of this ability came the study of logic, which is "the analysis and appraisal of arguments" (Gensler 2002: 305). Thus for logicians, definitions are important as the means of constructing and analyzing arguments. Logicians, like philosophers, are concerned with the nature of knowledge, but they take their curiosity a step further. Instead of focusing on what is knowledge and what isn't, those who study logic try to determine if a particular instance of knowledge is true or false. It is also important to mention that the distinction between definitions of things and definition of words is key to understanding a logicians' approach, for they deal with definitions of words. For example, when Gensler (2002: 305) explains the connection between construction of arguments and definitions, he draws on language as one of the ways to appraise arguments. He further explains that to a logician: "A definition is a rule of paraphrase intended to explain meaning. More precisely, a definition of a word or phrase is a rule saying how to eliminate this word or phrase in any sentence using it and produce a second sentence that means the same thing—the purpose of this being to explain or clarify the meaning of the word or phrase" (Gensler 2002: 308).

So while a philosopher often ponders the definitions of things, a logician deals in definitions of words. Thus a definition of the same object might be different depending which approach one takes. And not every approach is suitable for all defining activities. Robinson (1962: 161–165), for example, sees major problems with trying to define certain things. He sees the attempts to come up with *real* definitions (brief definitions that embody all essential knowledge about a thing) as not appropriate in all spheres of inquiry. While Robinson (1962: 163) agrees that it is possible to define some things by compressing "all the facts about *x* into a single short phrase," it becomes difficult to offer such definition for "life" or "individuality." Robison (1962: 163–164) claims that real definitions are possible only in certain disciplines, geometry or mathematics, for example. His concern seems to be particularly with one supposed requirement of a definition—that it is short. He writes, "There is a persistent tradition that a definition must be something brief, something certainly not more than a paragraph, and preferably not more than a sentence" (163).

The theory of definition when examined through the lens of philosophy and logic reveals that definitions are embodiments of knowledge. Calsamiglia and Van Dijk (2004: 385–386) demonstrate that this is true for definitions in popular science as well since explanatory strategies (including definitions) used in popular science texts correspond to general "categories of a basic knowledge schema for the representation of things."

In addition to reflecting ways in which human knowledge is constructed, definitions can also categorize knowledge as factual or subjective. This approach opens up a debate about the objectivity of definitions. Ancient Greeks did not accept definitions based on individual values and regarded them as inappropriate foundations for arguments. For example, in *Euthyphro*, Socrates adamantly objects to an individual interpretation of piety and insists on a general one (Plato 2011 [Cohen, Curd and Reeve eds.]: 135–152). Modern philosophers and logicians have no problem accepting definitions that are based on an individual's perception;

however, they do make a distinction based on the objective/subjective divide. Subjective definitions are called *stipulative* and assign meaning according to the definer's understanding (Harris and Hutton 2007: 3). Stipulative definitions represent the value side of what Schiappa (2003: 5) calls "the separation of fact and value," where "facts involve objective reality and values reflect subjective human preferences." Stipulative definitions are not the only ones to represent the realm of values. Other types of definitions to fall within this category may include *persuasive* definitions (definitions that influence an opinion [Copi 1972: 118–124]), *contextual* definitions (use of the term in a context that determines its meaning [Robinson 1962: 107]), and *lexical* definitions (definition of a word in common usage [Robinson 1962: 62]). To represent factual knowledge one may use *real* definitions ("a description of the objective nature of things" [Robinson 1962: 169–170]), *theoretical* definitions (definitions that "formulate a theoretically adequate characterization of the objects" [Copi 1972: 122]), and *ostensive* definitions (definitions that "make use of pointing or physical introduction" [Robinson 1962: 15]).

Studies of definitions in philosophy and logic are essential to understanding the history of development and analysis of definitions. However, these disciplines tend to take prescriptive approach to definitions—they focus on rules of formation and not on the examination of possible instances of definitions in texts. Linguistics, on the other hand, offers a more inclusive exploration of definitions.

Let us continue with the type of definition we have covered in detail so far—the A = B definition. As I mentioned before, this is a kind of definition you would find in a dictionary. Dictionary definitions are often regarded as traditional and authoritative. Robinson (1962: 36–37) for example, says that "the majority" of readers see dictionaries as preservers of "eternal and independent meaning" which is always "accurately stated." However, as Robinson (1962: 37) points out, dictionary definitions often disregard the fact that "the meanings of words cannot possibly be independent of man," and that a dictionary records only certain instances of word use by a particular social class. In his opinion, "a good dictionary would be only history" not an authority. Robinson (1962: 37) explains that dictionary definitions reflect human tendency for "approval and disapproval about the various ways of using words." Harris and Hutton (2007) take the argument further, suggesting that dictionary definitions are necessarily stipulative, reflecting the preferences and the understanding not of a specific social class, as Robinson suggests, but of the lexicographer—the person who puts a dictionary together or is in charge of the editorial team. It is also possible to see a dictionary as, in some ways, a product of a culture which produced it. But what about a dictionary of scientific terminology?

Bergenholtz and Tarp (1995:70) point out that scientific dictionaries are "independent of culture." They mean that scientific knowledge itself is not culture-dependent; it "does not change with country or language community" (Bergenholtz and Tarp 1995: 61). The separation of the knowledge from language, however, is arguably artificial, especially to a linguist. While a scientific dictionary will attempt to exclude culture-specific references, popular science will embrace them (see chapter 8).

A commonly observed way to define scientific terminology in a popularization would be to use a metaphor or an analogy, and such techniques will require a shared cultural knowledge of the author and the reader. As Toolan (1996: 60) explains, "Previous experience, including linguistic experience, is ... relevant to the interpretation of any particular utterance in context." Interpretation of metaphors or analogies involves transferring of "characteristic properties" from one object to another. Such "characteristic properties," however, "cannot denote ... the conventional meaning definition of an expression as supplied by a dictionary," but must be "associations" that are "remembered" by an individual from previous "form-context situations" (Toolan 1996: 65). If the reader and the author do not share a cultural background, they are likely to experience and recall different "form-context situations" when interpreting and creating metaphors, which might lead to confusion rather than a more approachable definition. This is why once we cross from dictionaries into popular science texts, culture independency becomes a less rigid line. In some cases, popular science authors take for granted certain cultural knowledge while using it as the basis for definitions. A good example would be the following definition from Greene (2011: 22):

> **constant negative curvature** ... means that if you view your reflection at any spot on a mirrored Pringles chip, the image will appear shrunken inward.

Knowing the shape of a Pringles chip becomes essential to understanding this definition. Thus scientific knowledge can be transmitted using highly culture-dependent language, which in turn makes the resulting definition culture dependent.

Definitions in popular science, while facing some of the same challenges as dictionary definitions, differ from them not only in the use of culture-dependent references, but also in the structural make up. For example, dictionaries, we have established, prefer to define by class and difference exclusively. This favoritism of a certain structure can be traced all the way back to the first edition of the *Oxford English Dictionary* (*OED*) published between 1889 and 1928 (Mugglestone 2000: 81). However, defining by delimiting (that is by class and difference) while producing concise definitions does not always help the reader understand the terminology fully. That is why popular science books do not rely on the dictionary type of definition exclusively (see chapter 8 for an in-depth discussion of definition types found in popular science). Not all dictionaries are constructed the same way, however.

Collins Cobuild English Learner's Dictionary (*CCED*) takes a different approach to definitions—"defining words in context, in full sentences which indicate context and lexicogrammatical patterning" (Moon 2009: 11). The reason to look at these definitions is that there is an important parallel between language learners and readers of popular science. A language leaner uses definitions in a dictionary to unlock the meaning of a text, and so does the reader of popular science—without definitions he/she would be lost inside even the simplest interpretation. As Myers (1990: 183) points out, "Definitions show how much scientific texts depend on certain terms," and popular science texts do not try to avoid scientific terminology but explain it.

Cobuild Dictionary defined by providing examples of possible word usage which included recognition of instances when a word had only subjective definitions. Barnbrook's (2002: 56) analysis of *CCELD* concludes that definitions with less "rigid structure" are more conductive to the understanding of the terminology, and "the definition sentences" where "headwords [the words being defined] are generally used as working units of language as well as being mentioned in the process of definition" provide a "more useful set of information." Barnbrook (2002) also explains that *CCELD* definitions differ from other dictionaries' by using a hinge—an "element that links" the subject with the describers. In the A = B scheme, the hinge is represented by the equal sign. Barnbrook (2002: 61) stresses the "crucial importance" of the hinge "since it specifies the nature of the relationship between the [subject and the describers], which is not always one of simple equality." The hinge is the element of a definition that allows it to be presented as an integrated sentence within a text since most of the time it provides the verb for the definition sentence. Consider the difference between an *Oxford Dictionary of Science* definition of gravity and that by Greene (2011):

> **gravity** The phenomenon associated with the gravitational force acting on any object that has mass and is situated within the earth's gravitational field [Daintith and Martin 2010: 368].

Before Einstein, **gravity** was a mysterious force that one body somehow exerted across space on another.... After Einstein, **gravity** was recognized as a distortion of the environment caused by one object and guiding the motion of others [Greene 2011: 14].

All of these definitions convey information about gravity, but in the dictionary definition, the subject is disjoined from the describers whereas in Greene's definition, the word "gravity" is a functioning part of the sentences. The difference comes from the absence of a hinge in the dictionary definition. Greene's (2011) definitions, on the other hand, use "was" and "was recognized as" to indicate the nature of the relationship between the parts of the subject and its describers. Greene's approach also reflects the historical development of scientific knowledge. It demonstrates the change in human understanding of gravity from "mysterious force" to "a distortion of the environment."

Ultimately, dictionary definitions offer two possible approaches: one is to produce a structurally rigid definition that attempts to be objective, and another is to construct a definition in context, which might lack some of the structural elements necessary for a prototypical definition yet will provide a more relatable description of the subject. I should note that the first approach, the one exemplified but the *OED* is considered traditional for dictionaries.

Out of the approaches suggested by philosophers, logicians, and lexicographers emerges what for the purposes of this book I consider a *prototypal definition*. It is a definition that uses A = B structure, where A represents the subject being defined, B signifies the describers (the class and difference), and the equal sign stands for the hinge, a verb connecting the definition to its subject and establishing the relationship between the subject and the definition. As I have demonstrated above, a hinge can be omitted in a dictionary definition, and if a prototypical definition appears in context, it will most likely have a hinge expressed by the verb "to be," as in:

The sphere *is* nature's easiest shape [du Sautoy 2011: 56].

A hinge, however, can be expressed by a variety of verbs. The most commonly used hinges include "to call," "to mean," and "to define," to give just a few examples. Another way to represent a hinge is by using a typographical mark (a dash or a colon, for example)—this is a means employed in dictionaries as well as in other texts. Darian (2003: 53–54) suggests that definitions found in science texts (and popular science is included in this group), often contain specific typographic marks. These are equal sign, colon, "pairs of commas," parentheses, "pairs of dashes," quotation marks, and italics.

The types of hinges mentioned above point to a relationship of equality between the subject and its definition that makes it possible to substitute a definition for the subject and vice versa. At the same time, equality is not the only type of a relationship between a subject and its definition that a hinge can express. Consider the following definition:

> A school *is supposed to be* a place where one acquires knowledge.

The hinge in this definition is expressing doubt rather than suggesting equality between "school" and "a place where one acquires knowledge." This example demonstrates that a hinge is not simply a connector but carries meaning and plays its own role in the understanding of a definition. In other words, a hinge can help introduce the definer's attitude into a definition, and a change in the hinge can easily convert a real definition (a definition commonly accepted as factual and objective) into a stipulative one (a definition expressing the definer's preferences).

Stipulative definitions receive a special treatment in linguistics, especially among integrationist linguists—those who look at language as acquiring meaning in context only, as opposed to recognizing that language has meaning outside of particular contexts where it is being used.

Harris and Hutton (2007: ix) see the integrationist theory as providing "a fresh approach" to definitions and asking "some questions about definition that have never been asked before." The new approach revolves around the theory's central idea that language without context is "indeterminate" and that "the meaning of a [linguistic] sign arises from the contextualized integration of particular activities in the communicational process" (Harris 2009: 3, 81). Since meaning comes about through specific communicational contexts only, the act of defining becomes vital when we exchange verbal or written messages (Harris and Hutton 2007: viii). The questions that the integrationist theory asks are closely related to the general thesis of indeterminacy—"if sense cannot be made of introducing a meaning by 'stipulation,'" why bother with "any other kind of definitional endeavor?" (Harris and Hutton 2007: ix).

For integrationists, stipulative definition "stands in the forefront of the whole topic" (Harris and Hutton 2007: ix). And while Harris and Hutton (2007) do not deny other kinds of definitions, the integrationist approach suggests that all definitions possess a degree of stipulation. Harris and Hutton (2007: 35) point out that "different individuals 'annex' different meanings to words," and resulting definitions are not entirely objective.

Applying this view to popular science, I can say that the definitions used in popularizations are not necessarily examples of set definitions, but simply one differential in the process of constructing meaning—the authors' understanding of terminology. The other aspect would be created by the readers once they process the information and start using the terminology applying their own understanding. Thus, if we follow the integrationist approach, popular science authors contribute not only to the dissemination of scientific knowledge but also to language usage. For instance, it is possible, if enough people read Greene's (2011) *Hidden Reality*, that defining "constant negative curvature" by referring to a Pringles chip will become a generally acceptable strategy. In fact, there is some evidence that something similar is already happening—however, not among the non-specialists but among popular science authors, who do not shy away from borrowing particularly apt metaphors from one another.

As Turney (2007: 2) notes, the use of figurative language is so ubiquitous to popular science that certain metaphors and analogies start to contribute to a pool of stock imagery that many authors "adopt and modify" for their own purposes. In contrast, new metaphors, according to Turney (2004b: 337), indicate that "there is not yet a widely accepted formula for describing ... novel" ideas. Turney (2004b: 343) suggests that the adaptation and modification of certain metaphors indicate the success of their originator.

A metaphor as a definition might seem unusual. At the same time, the integrationist view of metaphor suggests that it is not a deviation from "routine usage" but a normal linguistic process. Toolan (1996: 60) explains that "novelty in language usage" is not abnormal, and metaphors do not need a literal back-up in order to be understood. Many readers of popular science will form their understanding of the terminology based on metaphorical definitions alone without a supplement of prototypical, dictionary definitions.

According to Harris and Hutton (2007:18), "An integrationist approach recognizes definitions as being stipulative not necessarily on the basis of their author's expressed intentions ... but on the basis of the function they fulfill in integrating communication between users." In other words, a definition is stipulative not only because it expresses its author's understanding, but because it sets out certain guidelines that both the author and the reader/listener will use to communicate within a particular context. Thus definitions can be seen as constructed not by the author of a text exclusively but by the author and the reader or listener together. Allan Bell (1984) in his audience design framework attaches similar importance to the recipient of the message by suggesting that speakers/writers can be influenced by their audiences' linguistic preferences. Thus a stipulative definition may be both a reaction to the preferences of the people to whom it is presented and an attempt by the definer to influence the same people through his/her own understanding of the defined concept or object.

Harris and Hutton (2007: 10) propose that stipulative definitions play a role in "the modern proliferation of scientific terminology." Such definitions, they claim, can help create nomenclature for new theories. Harris and Hutton (2007: 11) suggest that "stipulating new names for new theoretical objects is ... an essential part of the advancement of science." Greene's (2011: 24–25) introduction of the term "dark energy" exemplifies this function of a stipulative definition: "To account for the more general possibility that the energy evolves, and to also emphasize that the energy does not give off light (exemplifying why it had for so long evaded detection) astronomers have coined a new term: dark energy. 'Dark' also describes well the many gaps in our understanding." This definition is of a theoretical entity since scientists are yet to determine the nature of dark energy. The name, as Greene (2011: 24–25) points out, reflects this lack of knowledge. Perhaps, once the object becomes less mysterious, its name will change. If this ever happens, the new definition will illustrate yet another function of stipulative definitions—"introducing a new scientific framework altogether ... claiming that all previous definitions were inadequate, i.e. based on fundamental misapprehensions" (Harris and Hutton 2007: 11).

Closely related to the idea that different people/communities might define in different ways is the rhetorical approach to definition analysis.

The rhetorical approach to definition theory is in some way similar to the logical approach. Both look at definitions in terms of their ability to construct and evaluate arguments, and both see definitions as contributing to the knowledge of social reality as opposed to philosophical approach, which sees definitions as describing physical reality. Rhetorical functions of definition also tie in with the integrationist view and Bell's (1984) framework for audience design in that all definitions are influenced by the definer and his/her audience. Edward Schiappa, a scholar of rhetoric based at MIT, explains that "a rhetorical analysis of definition ... investigates how people persuade other people to adopt and use certain definitions to the exclusion of others" (2003:4). In this way, Schiappa (2003), similarly to Harris and Hutton (2007) and Bell (1984), sees the process of definition construction as two-fold, including the definer and his/her audience, who ultimately decide on the success of a definition. Harris and Hutton (2007:3) point out that conflicting definitions used by parties in a court case can influence the outcome of the judicial process; thus illustrating that one definition was accepted in favor of another. "The difference between those definitions that are accepted and used and those that are not is a matter of persuasion" (Shiappa 2003: 3). This becomes useful when considering such definitions as this one of "consciousness": "I personally think that one of the problems has been the failure to clearly define consciousness and then a failure to quantify it. But if I were to venture a guess, I would theorize that consciousness consists of at least three basic components: 1. Sensing and recognizing the environment 2. Self-awareness 3. Planning for the future by setting goals and plans, that is, simulating the future and plotting strategy" (Kaku 2011: 97).

This example illustrates Shiappa's (2003) concept of a *novel* definition. A novel definition tries to establish a new use for a familiar term. In Schiappa's (2003: 11) own words, "Someone introducing a novel definition wants to change other people's understanding and linguistic behavior away from the conventional patterns." And this is exactly what Kaku (2011) tries to accomplish. He identifies a problem with existing definitions and suggests a solution in the form of his own novel definition. The problem that Kaku (2011) recognizes concerns what Shiappa (2003) calls a *mundane* definition. A mundane definition is "a definition that is used unproblematically by a particular discourse community" (Shiappa 2003: 29). In other words, it is a definition commonly accepted by a group of people; in this case scientists investigating consciousness. While Kaku (2011) does not supply a mundane definition of consciousness, he claims that it is unclear, "failure to clearly define consciousness." The fact that Kaku (2011) does not provide the definition he considers unsatisfactory establishes his novel definition as a more aggressive attempt to persuade the reader than it would have been otherwise.

The approaches I described above (from philosophy, logic, linguistics, and rhetoric) all attempt to characterize definitions as they relate to the understanding and, in some way, to the construction of reality. These approaches also share one similarity—they are theory driven and do not deal with many actual definitions as used in naturally-occurring texts. My study of definitions, which you will find in chapters 8 and 9, does not hinge on a single existing approach to definition analysis. Instead, I look at a variety of possible definitions used by popular science authors. This approach forces me to draw conclusions based on the specific data I find instead of trying to prove that certain frameworks could be observed.

The point of this chapter is to introduce to you some basic theoretical background on the three main research areas that are the foundation of the book—the study of narrative, presented discourse, and definitions.

I tried to keep the information given in this chapter fairly broad, with only occasional mentions of what all of this means to linguistic analysis of popular science. The rest of the book will fill that gap. Throughout the chapters that follow, I will refer to the ideas and terminology introduced here, and I trust that you will consider this chapter again should something seem unclear or should you need to brush up on the nomenclature.

The material from which I draw my conclusions comes from ten popular science bestsellers. The authors are equally divided between scientists and science writers (that is, people who do not have PhDs in science and have never practiced science but do write on scientific topics professionally). The books I used to collect my data are presented below, arranged by the authors' last names:

- Bill Bryson 2003, *A Short History of Nearly Everything*
- William Bynum 2012, *A Little History of Science*
- Sean Carroll 2012, *The Particle at the End of the Universe*
- Enrico Coen 2012, *Cells to Civilization*
- Marcus du Sautoy 2011, *The Number Mysteries*
- Timothy Ferris 1988, *Coming of Age in the Milky Way*
- Brian Greene 2011, *The Hidden Reality*
- Michio Kaku 2011, *Physics of the Future*
- Sam Kean 2012, *The Violinist's Thumb*
- Carl Zimmer 2011, *A Planet of Viruses*

These texts represent the following scientific disciplines: astronomy, chemistry, genetics, mathematics, medicine, physics, and virology. All of them are originally written in English, and most have been published between 2011 and 2012. However, the present study also incorporates two older texts: one from 2003 and one from 1988. The reason behind including older texts was to see whether examples taken from them fit within the analytical categories adopted primarily with the newer texts in mind. For example, one of the underlying claims that I make in this book is that popular science is directed towards the presentation of scientific issues through emotionally engaging mechanisms, especially dramatizing. Using older texts allows me to say that such a mode of presentation is not as novel as current research on popular science suggests. It was probably underreported by the analysts, who, as my overview of the field leads me to believe, became focused on the idea of the emotional connection with science fairly recently.

For the portions of the analysis that concern narratives, I examined 100 stories, with 50 narratives being taken from books written by scientists and 50 from books written by science writers. The narratives are, for the most part, between 200 and 500 words long, with the shortest being 150 words and the longest just over 4000 words.

These stories, be they tales of personal experience or of scientific discoveries, are not themselves the end products but rather brief excursions into personal history or the history of science. They function as guiding posts on a larger journey to uncovering the potentials of modern-day science. The stories often provide the reader with the necessary background on the fundamental laws and scientific principles or they demonstrate the human side of science.

For the analysis of presented discourse, I drew my data primarily from the narratives of discovery; however, I also examined discourse presentation types that occur outside discovery stories. The data from the narratives of discovery include 193 occurrences of discourse presentation of scientists. These occurrences are short stretches of discourse that are on average about 35 words long.

Overall, even though I do supply the frequency counts for the phenomena observed in narratives and in presented discourse of scientists, this is largely a qualitative study.

2

Personal Narratives

Perhaps the most influential study of narratives of personal experience to emerge from linguistics is the one done by William Labov in 1972. His earlier work on oral narratives was conducted together with Joshua Waletzky, and in 1966 they presented a paper titled "Narrative Analysis: Oral Versions of Personal Experience" at the Annual Spring Meeting of the American Ethnological Society in Philadelphia. The article published in the subsequent year was groundbreaking because it provided a systematic structural approach to oral narratives, suggesting that no matter the story or the background of the speaker, certain elements are common among all personal experience stories. In 1972, through a case study, Labov perfected this model. The original Labovian approach suggests that narratives of personal experience can be parsed into several sections that correspond to the following narrative elements: *Abstract* (an optional brief summary of a tale to come), *Orientation* (introduction of the characters and the setting), *Complication* (the main body), *Evaluation* (the speaker's assessment), *Result* (the resolution), and *Coda* (the element that transitions the speaker and the listener from the past events of a story to the present). Labov (1972: 369) suggests that a complete narrative contains these elements and presents them in the order listed above.

These structural elements are known as the *narrative macrostructure*. In addition to the narrative macrostructure comprised of the narrative elements, Labov and Waletzky (1967) and Labov (1972) outlined a microstructure which consists of narrative, free, and restricted clauses. The *narrative clauses* represent the events of the story in the chronological order (*the temporal sequence*) and cannot be rearranged without changing the meaning. For example, if we were to rearrange the verbs in the following sentence, we'd change the cause-effect relationship: *The girl saw a lion; the girl ran. The girl ran; the girl saw a lion.* The temporal sequence, therefore, is the narrative's skeleton, onto which details can be introduced using free clauses and restricted clauses. *Free clauses* are passages that are not part of the temporal sequence and can shift positions within a narrative. *Restricted clauses* are neither temporally ordered nor completely free and can be rearranged only within certain limits (Labov and Waletzky 1967: 22–23).

While the analysis of narrative microstructure can be interesting and useful for certain purposes, it is the narrative macrostructure examined by Labov (1972) in detail that propelled him to recognition (at least in the field of narratology). To this day, the model offers valuable insight into a variety of stories. Numerous studies proved its applicability to narratives other than those relating personal experiences. Labov's (1972) model (with modifications) has served as a springboard for structural analyses of narratives. The value of the model is in its universality; it is capable of exposing the underlying narrative structure of multiple text types. In fact, I will demonstrate its usefulness to the analysis of popular science narratives of discovery in the following chapter. For now, though, let's look at the personal narratives that occur in popular science books.

There are potentially a great number of personal narratives. If you ask a fisherman and a stockbroker to tell you a personal story, you will likely end up with two very different narratives. That is, the topics will be different, but the way they are arranged will be similar. When Labov (1972: 355–366) interviewed speakers of what he called "black English vernacular" in New York City, he asked his subjects to tell "'The Danger of Death'" stories. Here is how he explains the interview process: "At a certain point in the conversation the interviewer asks, 'Were you ever in a situation where you were in serious danger of being killed where you said to yourself—"This is it"?'"(Labov 1972: 355).

Of course, you will not expect narratives like these in a popular science book, but you should expect narratives since, as we have established in the previous chapter, stories make science more relatable and easier to understand. Personal stories are also excellent at explaining motivations and voicing the same questions that a reader might have. Here is an example from Bill Bryson's book *A Short History of Nearly Everything* (2003: 4–5):

> I was on a long flight across the Pacific, staring idly out the window at moonlit ocean, when it occurred to me with a certain uncomfortable forcefulness that I didn't know the first thing about the only planet I was ever going to live on. I had no idea, for example, why the oceans were salty but the Great Lakes weren't.... I didn't know if the oceans were growing more salty with time or less, and whether ocean salinity levels was something I should be concerned about or not. (I am very pleased to tell you that until the late 1970s scientists didn't know the answers to these questions either...) I didn't know what a proton was, or a protein, didn't know a quark from quasar.... I became

gripped by a quiet, unwonted urge to know a little about these matters.... So I decided that I would devote a portion of my life—three years, as it now turns out—to reading books and journals and finding saintly, patient experts prepared to answer a lot of outstandingly dumb questions.

Telling personal stories might seem like an awkward thing to do in a book about science—after all, professional scientists go to great lengths to remove themselves from the texts that they produce. Professional scientific prose is (on the surface) devoid of human emotion as much as possible. In the conclusion, I will address the relationship between professional and popular science in more detail; for now, let me explain why personal narratives make it into popular scientific books.

The answer lies in the idea that people are very self-centered, but not necessarily in a bad way. This means that we tend to process the world and our knowledge about it through personal experiences. As Schank (1990: 29) tells us, "The knowledge that people have about the world around them is really no more than the set of experiences that they have had." In other words, when Michio Kaku, Bill Bryson, or Brian Greene want to convince you (and, trust me, they do) that the scientific outlook on life is the most beneficial way for society to progress and learn, they share with you a personal story of what attracted them to science in the first place. This is how you get "How/Why I Became a Scientist" or "Why I Write about Science" narratives. Let us look at one more such story. Here is the opening of Brian Greene's bestseller *The Hidden Reality: Parallel Universes and the Deep Laws of the Cosmos*:

> If, when I was growing up, my room had been adorned with only a single mirror, my childhood daydreams might have been very different. But it had two. And each morning when I opened the closet to get my clothes, the one built into its door aligned with the one on the wall, creating a seemingly endless series of reflections of anything situated between them. It was mesmerizing. I delighted in seeing image after image populating the parallel glass planes, extending back as far as the eye could discern. All the reflections seemed to move in unison—but that, I knew, was a mere limitation of human perception; at a young age I had learned of light's finite speed. So in my mind's eye, I would watch the light's round-trip journeys. The bob of my head, the sweep of my arm silently echoed between the mirrors, each reflected image nudging the next. Sometimes I would imagine an irreverent me way down the line who refused to fall into place, disrupting the steady progression and creating a new reality that informed the ones that followed. During lulls at school, I would sometimes think about the light I had shed that morning, still endlessly bouncing between the mirrors, and I'd join one of my reflected selves, entering an imaginary parallel world constructed of light and driven by fantasy. To be sure, reflected images don't have minds of their own. But these youthful flights of fancy, with their imagined parallel realities, resonate with an increasingly prominent theme in modern science—the possibility of worlds lying beyond the one we know [Greene 2011: 3].

First of all, starting a book about science with a story is a smart move—narratives, as we learned in chapter 1, attract attention, and we respond to them better than we do to numbers or strings of facts. Opening with a personal narrative is doubly successful—it allows the author to establish a personal tone of the book right away. Greene (2011) takes a very complex subject and approaches it in a simple way: through the eyes of his younger self. At this point, the story is serving a dual purpose, it gives the reader an opportunity to relate to the author's personal experience, and it also lets the same reader know that the question of parallel universes (the subject of Greene's book) has been on the author's mind since he was a child. What was it that got Brian Greene thinking about parallel universes? The two mirrors in his room.

Let us examine each of the story's purposes in more detail. First of all, the relatability of this personal experience is relatively high. Many of us had played with mirrors in this way. Reading Greene's story reminded me of my own ventures into the mirror land. My journeys, however, were different. I preferred not to enter the alternate realities of the mirror corridor and positioned the mirrors in such a way that my own reflection was not part of the "parallel worlds." Rather, I was content with watching the endless replicas of the reflected objects in the room, imagining what it would be like to travel down that mysterious pathway. Perhaps, you have done something similar or know someone who has. And if you haven't had this particular experience, you can, using Greene's description, easily imagine it and even replicate it, should you want to amuse yourself.

The second purpose of this narrative—to establish the expertise of the author—is more subtle and might not be a shared experience between the reader and the author. Fairly early on, Greene (2011: 3) tells the reader, "at a young age I had learned of light's finite speed." This statement seemingly takes the magic out of the story and replaces it with science. It also tells the reader that the explanations for the phenomenon would not be fantastical or supernatural but scientific. Plus, the reader is to be impressed by the boy's knowledge. Later, in the story, Greene mentions "lulls at school," when his mind would wander back to the notion of parallel universes. These experiences are not meant to be shared. They are meant to reveal to the reader the author's early interest in science and to offer an initial proof (more will follow) that he knows what he is talking about—the question has occupied him for a long time.

There is a third point to this narrative as well. It is in the very first sentence. The most common household objects can contribute to a burgeoning interest in science. We don't know if the two mirrors in the young boy's room were an accident or a result of a deliberate placement by the parents who recognized their child's early abilities. Greene (2011:3) certainly makes it sound as if it was an accident: "If, when I was growing up, my room had been adorned with only a single mirror, my childhood daydreams might have been very different." That "If" makes all the difference. It suggests an alternative possibility. A parallel universe, if you will, where little Brian had only one mirror and never became a famous theoretical physicist. As Schank (1990: 36) explains, "The telling process, even in the relating of a firsthand experience, can be a highly inventive process. That is, the art of storytelling involves finding good ways to express one's experiences…. The entertainment factor exists in relating firsthand experiences just as it does in inventing stories. Nobody wants to listen to what happened to you today unless you can make what happened appear interesting." The possibility of this story turning out in a different way, that "If" at the beginning, certainly contributes to creating interest in Greene's narrative. At the same time, there are several other factors that inject interest and relevance into this narrative.

This story, and all other stories like it, are built by arranging certain elements in certain ways, and that particular arrangement is what creates interest. This is where Labov's (1972) order of narrative elements comes into play: the Abstract, the Orientation, the Evaluation, the Result, and the Coda.

We have already started examining the Abstract of Greene's (2011: 3) narrative when we discussed the hypothetical scenario that the first sentence presents. It is an unusual way to approach the Abstract—a brief summary of a story—but it certainly piques the interest. The summary is implied. Having come across this sentence, a reader would know that the story (at least partly) will be about childhood daydreams.

The Orientation—the setting of the scene—is next in Labov's sequence. In Greene's (2011: 3) narrative it is represented by: "But it had two. And each morning when I opened the closet to get my clothes, the one built into its door aligned with the one on the wall, creating a seemingly endless series of reflections of anything situated between them." Here we have the description of the room, or, at least, the relevant parts of it. We learn about the two mirrors—one on the wall and one in the closet. The setting is not exciting in itself, but without it, the rest of the story is impossible. A more elaborate and unusual surrounding would certainly contribute more to the interest of the story, but Greene (2011: 3) chooses not to use the Orientation in this way. This part of the narrative is dedicated to boosting the veracity of his story. Certainly, it is possible to arrange mirrors in the way described.

Schank's (1990: 36) claim that "firsthand stories are told because they relate information that is nonstandard in some way" not withstanding, not every portion of a narrative has to be devoted to something extraordinary. Greene (2011: 3) saves his nonstandard information for the Complication—the main body of the narrative, its plot, also the start of the temporal sequence.

The Complication, according to Labov (1972: 369), should follow the Orientation. However, Greene (2011: 3) takes a detour. "It was mesmerizing," he exclaims after describing the arrangement of mirrors in his room. This pause in the narrative action which began with "And each morning when I opened the closet to get my clothes" is not at all undesirable. It is known as the Evaluation—the author's reaction to what is happening in the story, his appraisal of the events being described to you.

Labov (1972: 369–370) says that the Evaluation is most often positioned after the Complication. That would be a logical place for the author to take a breath, look back at the actions or the events she has just described and offer her assessment. However, Labov (1972: 369) notes that "the evaluation of the narrative forms a secondary structure which is concentrated in the evaluation section but may be found in various forms throughout the narrative." That is, short evaluative remarks may be included at multiple points in the story in addition to the traditional Evaluation section following the main action of the narrative. It is also possible for the narrative to contain only short evaluative remarks throughout and no formal Evaluation. Labov (1972: 369) calls evaluative remarks that "penetrate the narrative" at multiple points *"waves of evaluation."*

In popular science narratives, the Evaluation usually serves as the mechanism for the author to show his enthusiasm and to try to inject interest into the story. This is what Greene (2011: 3) is doing as well. As a boy, he was "mesmerized" by the mirror corridor, and he hopes the reader would also get caught up in this excitement.

One important point about the Evaluation that Labov (1972: 369) makes is that it "forms a secondary structure" of a personal experience story. This means that the Evaluation itself is not absolutely necessary to a narrative; it is not part of the temporal sequence. Yet, while some personal experience narratives would be missing Abstracts, none would skip the Evaluation. A secondary structure that the Evaluation creates is actually invaluable. It allows for the emotional connection between the author and the reader (or the speaker and the listener) as both share in the emotion expressed through it.

Popular science narratives go one step further and extend the secondary structure of certain narratives by introducing one more element that is outside the temporal sequence yet very important to the story—the Explanation. I will discuss it in full detail in the following chapter that is devoted to popular science narratives of discovery—the types of stories that are more likely to contain this element. The Explanation is not part of the sequence of narrative elements introduced by Labov (1972), but it is present in almost all popular science narratives.

In the narrative we are examining now, the Explanation interrupts the Complication. It is the portion of the story that reveals Greene's (2011: 3) early interest in physics, "but that, I knew, was a mere limitation of human perception; at a young age I had learned of light's finite speed." The same sentence that let's the magic out of the story and ushers in the science.

The Complication itself takes up the majority of the narrative and relates young Brian Greene's movements in front of the mirror and the behavior of the multiple reflections that he saw. Traditionally, as the temporal sequence of a narrative, the Complication would contain the most action verbs compared to any other section. In our narrative, we have "seemed to move," "would watch," "echoed," "nudging," "would imagine," "refused," "disrupting," "creating," and "informed." This is almost double the number of verbs in any other section of the story.

The preponderance of verbs in Complications aside, they are notable because through these verbs it is possible to trace the main plotline of the story. They indicate the most important action nodes. Notice also the difference in the verbs that describe the author's perceptions and the verbs that denote the actions of his reflection. The first group of verbs ("seemed to move," "would watch," "would imagine") all express tentativeness, as if to continue with the hypothetical tone set by the Abstract, that mysterious "If" at the beginning of the story. There is, however, nothing tentative about the actions of the author's reflected self. There is no hedging, no "would" or "seemed" preceding the verbs to indicate a degree of uncertainty.

Isn't it interesting that Greene (2011: 3) would choose to relate the world of reflections in definite terms while introducing a note of hypothesis into the description of his real self? It is definitely a way to relate information in a "nonstandard" way—a way, according to Schank (1990: 36) to create interest in the story.

From the Complication we move onto the Result, where we learn what a long-lasting impression the multiple reflections made on young Brian Greene. At the end of the story, we find out about those "lulls at school" when he would think of physics. Greene (2011:3) is not explicitly stating this, but we know from the information given earlier that he understood the inner workings of those multiple reflections and approached them from a scientific point of view rather than in a purely fantastical manner. Thus, the Result of this narrative solidifies the author's interest in parallel universes from an early age and points to this childhood fascination as a contributing factor in his career choice.

Yet the story is not completely over. Now comes the time to connect this very personal prelude to the larger subject of the book—parallel universes. The last two sentences of the narrative represent the Coda—a narrative element that transitions the reader or the listener out of the narrative into the broad subject of a text or conversation. The Coda usually marks the end of the temporal sequence. It is a place where the reader is most likely to observe a shift from the past tense of the narrative into the present tense of the larger text.

In Greene's (2011: 3) story, the Coda transitions the reader not only from the past to the present but also out of the hypothetical world of the reflections ("would imagine," "would think," etc.) into a more solid reality represented by the present simple of the verb "resonate."

It might appear unusual that a narrative structure designed to explicate stories of near-death experiences of the speakers of African American Vernacular in New York City in the 1970s corresponds so closely to the narrative makeup of a popular science narrative published in 2011. However, such is the power of narrative that this is not surprising. Stories following essentially the same structure could be found across time and genres. A number of studies have confirmed that, in essence, Labov's (1972) structural elements—first determined through analysis of oral personal experience stories—can be observed in narratives told by small children (Peterson and McCabe 1981; Berman 1997), in autobiographical stories of adults (Baerger and McAdam 1999), and in literary narratives (Fleischman 1997), to give just a few examples.

Once a successful story structure is created, it tends to repeat itself—we respond favorably to the stories we like, thus encouraging proliferation of particular narrative types. For example, a heroic romance story originated in the English language with *Beowulf* is popular to this day. In fact, Michael Hoey (2001: 121–169) singles out a handful of what he calls "culturally popular patterns of organization" for stories composed in Western societies. Hoey's (2001) patterns of text organization include Problem-Solution, Gap in Knowledge-Filling, Goal-Achievement, Opportunity-Taking, and Desire Arousal-Fulfillment patterns. The names of these story patterns indicate how the action will progress in each one.

Out of the five patterns, Hoey (2001: 146, 162) singles out two as most often found in popular science: the Goal-Achievement pattern and the Gap in Knowledge-Filling pattern. Hoey (2001: 123) also considers the Problem-Solution pattern "the most common" overall. My data confirm one portion of his conclusions: the Gap in Knowledge-Filling pattern appears to be the most popular pattern in the popular science texts I analyzed. At the same time, the Goal-Achievement pattern did not occur as often as it should have according to Hoey's (2001) determination. The Problem-Solution pattern is the second most frequent pattern of story organization in popular science books, according to my observations.

It is only natural that the Gap in Knowledge pattern will prevail in the stories about science. After all, science is all about expanding our knowledge. So when in the next chapter we look at the narratives of scientific discovery, the Gap in Knowledge pattern will receive a lot of attention. It is more unusual to find a personal experience narrative that is structured as a Gap in Knowledge story outside of popular science (Bryson's story is an example). I say this is unusual because, in general, the Problem-Solution pattern is more popular, and certainly stories about personal problems abound in our society. In fact, think of the most recent fictional story you read or the most resent news report—they were likely stories of people experiencing and overcoming problems. We report and invent such stories with greater frequency because they function as learning mechanisms: the readers or listeners can co-experience the behavior necessary in certain difficult situation and then, if necessary, repeat it themselves, thus ensuring a greater chance of success. As Boyd (2008) puts it, "One way ... to approach literature (and art in general) is to view it as a behavior in evolutionary terms." He continues to explain why certain types of stories are more popular than others, using survival of the species as an underlying mechanism: "Although house buying has become a stressful preoccupation in modern life, we have no genre of real-estate novels. But we do have stories about romantic love. An evolutionist can note the significance of reproduction and survival in the transmission of genes and the evolution of species. This can explain why, over countless generations, our emotions have been designed to respond so intensely to love and death, and why romance stories so often focus on finding love or that thrillers, mysteries, and adventure tales focus on avoiding death." Love stories, thrillers, and mysteries all follow the Problem-Solution pattern. So whether we look at these narrative structures from the point of view of linguistics or evolution, the result is the same: stories about solving problems are the most common narratives.

Does science not solve problems? Of course, it does. However, popular science books are devoted largely to explorations of intriguing scientific issues and puzzling questions, not primarily to practical applications of the solutions though Problem-Solution stories are certainly present and represent the second most often-used pattern.

To return to personal narratives, let me share an example that illustrates both Labov's (1972) arrangement of narrative elements and Hoey's (2001) Gap in Knowledge-Filling story pattern. It comes from Michio Kaku's (2011) book *Physics of the Future: How Science Will Shape Human Destiny and Our Daily Lives by the Year 2100*. Like Brian Greene, Kaku is a theoretical physicist, and just like Greene's (2011) personal story, Kaku's (2011: 1–2) narrative is in the opening paragraphs of the book:

> When I was eight years old, I remember all the teachers buzzing with the latest news that a great scientist had just died. That night, the newspapers printed a picture of his office, with an unfinished manuscript on his desk. The caption read that the greatest scientist of our era could not finish his greatest masterpiece. What, I asked myself, could be so difficult that such a great scientist could not finish it? What could possibly be that complicated and that important? To me, eventually this became more fascinating than any murder mystery, more intriguing than any adventure story. I had to know what was in that unfinished manuscript. Later, I found out that the name of this scientist was Albert Einstein and the unfinished manuscript was to be his crowning achievement, his attempt to create a "theory of everything," an equation, perhaps no more than one inch wide, that would unlock the secrets of the universe and perhaps allow him to "read the mind of God."

This is a perfect example of a Gap in Knowledge narrative. There are several signs in the story that point to that particular pattern.

Hoey (1983, 2001) calls such signs *lexical signals*. Lexical signals are words that help the reader understand the relationships between sentences or groups of sentences in a given text. According to Hoey (1983: 63), "Lexical signaling can take the form of a sentence, clause or phrase and incorporates either one or more typical signals or an evaluation." Hoey (2001) calls the first signal of a pattern a *trigger*. The trigger determines the type of pattern.

Hoey (2001: 124) builds the idea of a lexical signal on the proposition that certain words trigger certain expectations of how a story will proceed; therefore, each pattern is triggered by its own lexical signals. For example, Hoey (2001: 124–125) proposes that such "negative evaluation items" as "problem" or "unable" are signals of the Problem-Solution pattern. For the Goal-Achievement pattern, Hoey (2001: 146–147) identifies several possible triggers including, "want to," "would like to," "aim," "means," etc.; any words that suggest setting of a goal would function as triggers of this pattern.

The Gap in Knowledge-Filling pattern is triggered by phrases that suggest a lack of knowledge, for example, "did not know" or "was puzzled." In Kaku's (2011: 1–2) narrative the trigger is in the form of a question, "What, I asked myself, could be so difficult that such a great scientist could not finish it?" Another question that follows the trigger works as an additional pattern signal and reaffirms the Gap in Knowledge setup for the story.

Having read these questions, you probably knew that by the end of the story the author would give you the answers, which of course Kaku (2011: 1–2) does. Had he posed these questions but left them unattended, the story would not work. It would not only break the readers' expectations, but it would also cease to be a narrative because it would lose the structural makeup necessary to be considered as such. The filling of the Gap in Knowledge is signaled by the phrase "Later, I found out," which appears, as it should, towards the end of the story.

It is possible to superimpose Labov's (1972) model onto Hoey's (2001) narrative patterns. In fact, the two analytical frameworks complement each other and help highlight a crucial feature of popular science narratives, which I will discuss in chapter 3 since it appears more prominently in the narratives of discovery and not in personal narratives.

Both narratives of discovery and personal narratives when analyzed using the combination of Labov's (1972) and Hoey's (2001) models reveal that the elements called the Complication and the Result can be further subdivided into several subtypes depending on the pattern of the story. Thus, if we examine Kaku's story using both Labov's (1972) and Hoey's (2001) ideas about narrative structure, we will find that the Complication is expressed by the author's lack of knowledge: "What, I asked myself, could be so difficult that such a great scientist could not finish it? What could possibly be that complicated and that important? To me, eventually this became more fascinating than any murder mystery, more intriguing than any adventure story. I had to know what was in that unfinished manuscript." The complicating action in the Gap in Knowledge narratives can be labeled *gap-in-knowledge Complication*—a kind of complication action where the author admits to not knowing something and wanting to remedy the situation.

In turn, the Result of Kaku's story also forms a subtype called the *filling Result*—it signals a particular type of resolution; that is, learning something new, finding out the answer to the questions posed by the Complication. In Kaku's (2011: 1–2) narrative, the Result is expressed by "Later, I found out that the name of this scientist was Albert Einstein and the unfinished manuscript was to be his crowning achievement, his attempt to create a 'theory of everything,' an equation, perhaps no more than one inch wide, that would unlock the secrets of the universe and perhaps allow him to 'read the mind of God.'" The combined approach confirms the ubiquitous nature of Labov's (1972) narrative structure, but it also demonstrates a certain flexibility to his model. While each narrative will contain the elements outlined by Labov, their functions will adjust depending on the type of story being told. For instance, it is impossible to have a story with the Complication that points to a setting of a goal (as in the Goal-Achievement pattern) and the Result that talks about a solution to a problem (as in the Problem-Solution pattern).

This does not mean, however, that it is impossible to construct different stories out of the same material. Greene (2011: 73) also addresses Einstein's death and his unfinished theory. His narrative follows the Goal-Achievement pattern (or rather lack of achievement) and is an example of a subgroup of discovery stories called narratives of failed discoveries. I address them in detail in the next chapter. Here is an excerpt from Greene's (2011: 73) narrative that directly discusses Einstein's death: "His personal secretary and gatekeeper, Helen Dukas, was with Einstein at the Princeton Hospital during his penultimate day, April 17, 1955. She recounts how Einstein, bedridden but feeling a little stronger, asked for the pages of equations on which he had been endlessly manipulating mathematical symbols in the fading hope that the unified field theory would materialize. Einstein didn't rise with the morning sun. His final scribblings shed no further light on unification." Not being a personal narrative, this story is quite different and does not have the optimistic spin that Kaku (2011: 1–2) puts on it: yes, Einstein is dead, and his work is unfinished, but that is what drew me to science. Personal narratives found in popular science tend to have positive and optimistic messages; their goal is to share the excitement of discovering science, not to make the reader privy to disappointments.

Even personal stories that do not necessarily discuss the first interest in science, narratives of personal research, for example, are likely to contain positive outcomes.

Of course, these research narratives are extremely simplified and shorn of all but the very basic technical details, if any appear at all. As a result, they present scientific research as more of an intellectual than a hands-on activity (this, incidentally, is also true for the narratives of discovery). Let's look at an example of a personal research narrative from Enrico Coen, a plant molecular geneticist:

> I came across an insightful essay written in 1990 by computer scientist Christoph von der Malsburg, in which he identified three basic principles that were common to self-organizing systems. His main emphasis was on brain function and development, but he also noted the importance of these principles for evolution. Reading his ideas led me to take a fresh look at the foundations of biological thinking, particularly in the light of what has been discovered in recent years. I proceeded to strip down the theories in each area of biology to their base essentials and then reexamine them afresh, looking at what they might have in common.... The surprising result, though, was that the commonalities that emerged were not superficial, but went to the very heart of each process.... I had arrived at a deeper and more unified way of seeing living transformation: different manifestations of a common creative recipe [Coen 2012: 10].

As you can see, there is not even a mention of laboratory work or scientific experiments. This is an extreme example of simplification. Biology is, after all, an applied science—it is impossible to practice it by just thinking, unlike theoretical physics, for example. So it would be logical to expect some details of lab work or at least references to experiments conducted by other researchers that Coen drew upon. Yet none are present. All we have is the description of his thinking process and a positive outcome of the research.

In the introduction, I mentioned popular science's preference for stories that show what Harré (1994) labeled "a smiling face of science." These are stories that always have positive results. For instance, Coen (2012: 10) set out to research commonalities in creative processes that take place in various living organisms and achieved satisfactory results, which are the subject of his book *Cells to Civilization: The Principles of Change That Shape Life*.

Much more rarely, narratives of personal research include negative results. However, these are not stories of ineptitude or failure—they are narratives of uncharted territories, of science yet unknown, and of brave researchers who struggle to unveil new laws of nature. Greene (2011: 91) supplies a representative example. I introduce it here divided into the basic narrative elements:

> Orientation:
> When I started working on string theory, back in the mid–1980s, there were only a handful of known Calabi-Yau shapes, so one could imagine studying each, looking for a match to known physics. My doctoral dissertation was one of the earliest steps in this direction.
>
> Complication:
> A few years later, when I was a postdoctoral fellow ... the number of Calabi-Yau shapes had grown to a few thousand, which presented more of a challenge to exhaustive analysis.... As time passed, however, the pages of the Calabi-Yau catalog continued to multiply ... they have now grown more numerous than grains of sand on a beach.... To analyze mathematically each possibility for the extra dimensions is out of the question.

Result:
String theorists have therefore continued the search for a mathematical directive from the theory that might single out a particular Calabi-Yau shape as "the one." To date, no one has succeeded.

Note the point of view shift that happens between the Complication and the Result sections: a narrative that begins and develops as a personal story (the use of pronouns "I" and "my") concludes with a general statement about the scientific community—"no one has succeeded." The negative Result is not a personal failure of Brian Greene but a collective lack of success. That is not to say that the author is avoiding taking the responsibility; by implementing this kind of shift in perspective, he is magnifying the nature of the problem—it is not a personal research goal that did not find support; it is the challenge that current scientific knowledge cannot overcome.

Such a narrative device allows the author to introduce a negative outcome and at the same time cushion it. This is what popular science authors want you to take away from such stories: science is not all powerful; it fails only (and notice that *only*) in face of insurmountable challenges, and even those failures are temporary. Greene (2011: 91) makes it a point that the lack of solution is "to date"; the answers will inevitably come in the future. In effect, this is still a smiling-face narrative.

The tendencies of scientific narratives that come through to the reader in personal stories (the lack of experimental descriptions, the focus on positive outcomes) become more pronounced in the narratives of discovery.

Personal stories, while not abundant in popular science, appear to be a staple of the genre, and many authors choose to share details of their personal lives with their readers.

The primary role of one type of personal experience story in popular science books is to draw the readers' attention to the authors as personalities and as people who were not always scientists. Such narratives relate an event or a series of events that boosted the authors' interest in scientific research. Their placement on the opening pages of the books suggests an almost autobiographical approach to the content.

Another group of personal narratives deal with the authors' research projects, sometimes introducing details of their PhD dissertations. These narratives tend to appear much later in the books, once the main subject has been well established and all preliminary details have been addressed. Their placement confirms the chronological progression of the overarching narrative of the books.

In between the opening narratives that trace the authors' road to science and the narratives of personal research, one will find narratives of scientific discoveries. These stories are dispersed throughout the books and most often cover the background material necessary for the discussion of the main subject. These narratives are the subject of the next chapter.

3

Narratives of Discovery
Explanation Made Easy

Turney (2004b: 333) surveys the kinds of narratives employed in popular science and suggests that they all serve one purpose—to explain or "translate" science into laymen's terms. To understand scientific stories in this way is to assign to them an overarching scaffolding function. Such an approach is not without support since, as I pointed out in chapter 1, our brains are predisposed to information that comes in the form of stories. Nowhere is the scaffolding role of popular science narratives more apparent than in the stories of scientific discoveries.

Narratives of discovery tell stories of how a scientist (or a group of scientists) comes up with new ideas, concepts, and breakthroughs. These are predominantly narratives about people (Gilbert and Mulkay [1984] note the importance of the human factor in science); however, they are rarely presented as personal experience stories even though they do describe the attitudes and experiences of the scientists involved. As third-person narratives, the narratives of discovery convey not only the experiences of making discoveries but also interject the authors' view and evaluation of the scientific discipline described and his/her attitudes and evaluations of the discovery (Pilkington 2017: 10).

The explanatory powers of narratives of discovery are so pronounced that when the stories are subjected to a Labov-style analysis (see chapter 2), they reveal additional narrative elements devoted entirely to anticipating the readers' needs and providing quick and easy explanations without disrupting the flow of the story.

In addition to the elements of the Abstract, the Orientation, the Complication, the Evaluation, the Result, and the Coda outlined in Labov (1972), I have identified an additional element, which I shall call "the

Explanation." The Explanation is part of what Labov (1972: 369) considers a "secondary structure" of a narrative, and together with the Evaluation it supplies information that is not part of the narrative core (or temporal sequence) but is, in many cases, essential for the processing of the narrative by the reader/hearer.

In the popular science narratives of discovery, the Explanation is a scaffolding mechanism used by the authors to clarify and explicate scientific concepts to lay readers. It is common to regard explanation as a type of text rather than as a narrative element. Turney (2004b) and Herman (2009) exemplify that view. Looking at explanations in scientific discourse, Herman (2009: 100–104) suggests that either explanation is a text type separate from narrative, or both explanation and narrative constitute a combined text type of explanatory narrative. Nowhere in current literature is there a suggestion to treat explanation as an element of a scientific (a popular scientific) narrative. It is much more common to assume as Turney (2004b) and Herman (2009) do that *all* of a story is devoted to explaining.

Having looked at narratives in popular science in some detail, I can say with confidence that not all of them have explanation as their only goal and that specific parts of such narratives (as opposed to the whole narratives) are explanatory. For example, as I will demonstrate in the following chapter, the narratives of discovery deliver not only information about scientific breakthroughs but also present a specific view on how science works. In these narratives, the explanatory function may be supplemented by an ideological one, representing a certain view of science. In light of the current research on narrative and its explanatory abilities (see chapter 1), it might be somewhat counterintuitive to argue that not all portions of a story are devoted to explanation. At the same time, structural analysis applying Labov's (1972) model clearly underscores *separate functions of each element in a narrative*. The Explanation in the narratives of discovery is usually confined to a particular section and does not dominate the story. Consider the following example taken from Marcus du Sautoy's book *The Number Mysteries: A Mathematical Odyssey through Everyday Life*:

"Galileo and Pendulum"

1. Abstract
   Why Are Pendulums Not As Predictable As They First Appear?
2. Orientation
   It was Galileo, the master of using math to make predictions, who first unlocked the secret of what makes a pendulum tick. The story goes that when he was 17, he was attending Mass at the cathedral in Pisa.

3. Complication
In a moment of boredom, he stared up at the ceiling, and his eyes fell on a chandelier that was swinging gently in the breeze blowing through the building. Galileo decided to time how long it took the chandelier to swing from side to side. He didn't have a watch (they hadn't been invented yet), so he used his pulse to keep track of the swing.

4.a. Result
The great discovery he made was that the time the chandelier took to complete one swing did not seem to depend on the size of the swing.

5. Explanation
In other words, the time of the swing essentially doesn't change if you increase or decrease the angle of swing. (I put the word essentially in there to indicate that if we dig a little deeper, things get slightly more complicated.)

4.b. Result
When the wind blew harder, the chandelier swung through a larger arc but took the same time to swing as when the wind dropped and the chandelier was hardly moving at all.

6. Evaluation
This was an important discovery and resulted in the swinging pendulum being used to record the passage of time.

7. Coda
But what does the time of the swing depend on, and can we predict whether and how the swing will change if the weight is increased or if the pendulum is made longer? [du Sautoy 2011: 226].

The Explanation is marked by the shift in tense from the past simple in the main body of the narrative to the present simple in the Explanation. It is clearly a section of a narrative that is not part of the temporal sequence. In fact, it interrupts the Result—the final element of the temporal sequence. While the last sentence of the Orientation, all of the Complication, and the Result recount the discovery, the Explanation clarifies the scientific concept uncovered by Galileo.

The analysis of the "Galileo and the Pendulum" according to Labov's model demonstrates that the narrative as a whole contributes to the readers' understanding, but the Explanation is the place where that understanding is the most salient. In other words, the Explanation focuses the information presented throughout the narrative.

Functionally and structurally, the Explanation is similar to Labov's (1972) Evaluation in a sense that in many narratives it works as a suspending device for the story and could be excluded without changing the flow of events. At the same time, the Explanation is closely interwoven into the fabric of a narrative and its exclusion may impact understanding. To popular science narratives of discovery, the Explanation might be as essential as the Evaluation is to narratives of personal experience. Labov suggests that without the Evaluation, such narratives lose their purpose. Moreover, Labov

(1972: 372) demonstrates that under certain circumstances, the Evaluation in the narratives of personal experience is essential in order to elicit the right emotional response from the audience and to keep them interested. The absence of the Evaluation in such cases will cause the narrative to break down because it will become unclear to the audience. The Explanation works very similarly. Without it, narratives of discovery may lose much of their meaning since a non-professional reader might not be equipped to understand the very subject matter of a discovery.

The double secondary structure of narratives of discovery created by the Evaluation and the Explanation performs a supportive role. It helps the reader construct a full picture of a scientific breakthrough by explaining the discovery and by outlining its importance within the bounds of the scientific community and beyond.

While the Explanation is usually a specific section of a narrative, it is possible for it to permeate the other elements and thus create what I call "explanatory waves." My choice of this label is influenced by Labov's (1972: 369) term "waves of evaluation." He writes, "The evaluation of the narrative forms a secondary structure which is concentrated in the evaluation section but may be found in various forms throughout the narrative" (Labov 1972: 369).

Explanatory waves usually manifest as short definitions or asides found within the elements representing the temporal sequence. For example, in the following narrative about the shapes of viruses, du Sautoy (2011: 69) includes not only the general Explanation for the discovery but also a brief definition of the icosahedron, creating an explanatory wave within the Result [all the explanatory material is bolded including the Explanation section itself and the explanatory wave in the Result].

"Shapes of Viruses"

Orientation
Having cracked the structure of DNA, Francis Crick and James Watson, along with Donald Caspar and Aaron Klug, turned their attention to what the two-dimensional pictures from X-ray diffraction could reveal about viruses.

Complication
To their surprise, they found shapes full of symmetry. The first images showed dots arranged in triangles, which implied that the virus had a three-dimensional shape that could be spun by a third of a turn and look the same. When the biologists looked in the mathematicians' cabinet of shapes, it was the Platonic solids that seemed the best candidates for the form of these viruses. The problem was that all five of Plato's shapes had an axis about which you could spin the shape by a third of a turn so that all the faces realign. It was only when the biologists obtained another diffraction image that they got a view that enabled them to pin down the shapes of these viruses more precisely.

Result
Suddenly, dots arranged in pentagons appeared, and that allowed them to home in on one of the more interesting of Plato's dice: **the icosahedron—the shape made of 20 triangles with five triangles meeting at each point.**

Explanation
Viruses like symmetrical shapes because symmetry provides a very simple means for them to multiply, and that is what makes viral diseases so infectious—in fact, that's what virulent means. Traditionally, symmetry has been something people have found aesthetically appealing, whether it is seen in a diamond, a flower, or the face of a supermodel. But symmetry isn't always so desirable. Some of the deadliest viruses in the biological books—from influenza to herpes, from polio to the AIDS virus—are constructed using the shape of an icosahedron.

In addition to creating explanatory waves, the Explanation can be presented as a series of elements. These separate sections are different from explanatory waves because they interrupt the temporal sequence by creating a tense shift; explanatory waves are located within the temporal sequence and do not initiate tense shifts. In the narrative "Bubbles Fusing Together," du Sautoy (2011: 70–74) interrupts the story three times in order to insert explanations; notice the tense shift from the past to the present at the beginning of each explanatory section:

"Bubbles Fusing Together"

1.a. Abstract
The first proof that the fused bubbles couldn't be bettered was announced in 1995. Although mathematicians don't really like asking for help from a computer (because that doesn't appeal to their sense of elegance and beauty), they needed one to check through the extensive numerical calculations that were involved in their proof. Five years later, a pencil-and-paper proof of the double-bubble conjecture was announced. It actually proved a more general conjecture:

2.a. Explanation
if the bubbles do not enclose the same volume, but rather one is smaller than the other, then the bubbles fuse together so that the wall between the bubbles is no longer flat but bent into the small bubble. The wall is part of a third sphere and meets the two spherical bubbles in such a way that the three soap films have angles of 120 degrees between them.

3. Orientation
In fact, this 120-degree property turns out to be a general rule for the way soap bubbles fuse together. It was first discovered by Belgian scientist Joseph Plateau, who was born in 1801.

4.a. Complication
While he was doing research into the effect of light on the eye, he stared at the sun for half a minute, and by the age of 40, he was blind. Then, with the help of relatives and colleagues, he switched to investigating the shape of bubbles. Plateau began by dipping wire frames into bubble mixture and examining the different shapes that appeared.

2.b. Explanation
For example, when you dip a wire frame in the shape of a cube into the mixture, you get 13 walls that meet at a square in the middle. This "square," however, isn't quite a square—the edges bulge out.

4.b. Complication
As Plateau explored the various shapes that appeared in different wire frames, he began to formulate a set of rules for how bubbles join together.

5.a. Result
The first rule was that soap films always meet in threes at an angle of 120 degrees. The edge formed by these three walls is called a Plateau border in his honor. The second rule was about the way these borders can meet.

2.c. Explanation
Plateau borders meet in fours at an angle of about 109.47 degrees ($\cos^{-1} 1/3$, to be precise). If you take a tetrahedron and draw lines from the four vertices to the center, you get the configuration of the four Plateau borders in foam. So the edges in the bulging square at the center of the cube wire frame actually meet at 109.47 degrees.

5.b. Result
Any bubble that didn't satisfy Plateau's rules was believed to be unstable and would therefore collapse to a stable configuration that did satisfy these rules.
It was not until 1976 that Jean Taylor finally proved that the shape of bubbles in foam had to satisfy the rules laid down by Plateau.

6. Coda
Their work tells us how the bubbles connect together, but what about the actual shapes of the bubbles in foam? Because bubbles are lazy, the way to the answer is to find the shapes that enclose a given amount of air in each bubble in the foam while minimizing the surface area of soap film.

As a scaffolding mechanism, the Explanation is also the most interactive element of a narrative of discovery, and it is usually here that communication between the author and the reader takes place. By stepping out of the temporally ordered events of the discovery story, the author is able to address the reader directly and ask him/her to engage the imagination to help comprehend the events described. Notice the interpersonal pronoun "you" in the explanatory sections of the "Bubbles Fusing Together" narrative above. du Sautoy is not simply addressing the reader but is using the pronoun to draw him/her into the hypothetical activities used to create the Explanation: "when *you* dip a wire frame in the shape of a cube into the mixture, *you* get 13 walls..." or "If *you* take a tetrahedron and draw lines from the four vertices to the center, *you* get the configuration of the four Plateau borders in foam."

In the "Duality" narrative, Greene (2011: 109–111) takes this technique one step further and uses the interpersonal pronoun "you" [bolded in the example] while relating his personal experience. By doing so he creates an illusion of the reader seeing the same things as he did. Greene (2011: 109–

111) thus transforms his own observations into an experience of the reader, helping him/her connect with the material on a personal level:

> I recently encountered a splendid graphic that from close up looks like Albert Einstein, with a bit more distance becomes ambiguous, and from far away resolves into Marilyn Monroe.... If **you** saw only the images that come into focus at the two extremes, **you**'d have every reason to think **you** were looking at two separate pictures. But if **you** steadily examine the image through the range of intermediate distances, **you** unexpectedly find that Einstein and Monroe are aspects of a single portrait. Similarly, an examination of two string theories, in the extreme case when each has a small coupling, reveals that they're as different as Albert and Marilyn. If **you** stopped there, as for years string theorists did, **you**'d conclude that **you** were studying two separate theories. But if **you** examine the theories as their couplings are varied over the range of intermediate values, **you** find that, like Albert turning into Marilyn, each gradually morphs into the other.

Another way to engage the reader is to use imperatives or directives within the Explanation. Greene's (2011: 84–85) narrative "Kaluza-Klein Theory" provides an example (imperative verbs in bold):

> To picture this, **think** of a common drinking straw. But for the purpose at hand, **make** it decidedly uncommon by imagining it as thin as usual but as tall as the Empire State Building.... Now **imagine** viewing the tall straw from across the Hudson River.... For another visualization, **think** of a huge carpet blanketing Utah's salt flats.

Overall, the major function of the Explanation is to create scaffolding for the lay reader. Together with the Evaluation, it forms the secondary structure of the narratives of discovery by suspending the main events to help the reader understand more fully the narrative and its significance.

The Explanation is not part of Labov and Waletzky's (1967) or Labov's (1972) macrostructure of a narrative. However, it is an important feature of popular science narratives of discovery which is revealed through the application of their model. The presence of the Explanation demonstrates a structural complexity of popular science narratives of discovery. Not only does it supply an additional element for the narrative structure, but it also reveals the possible re-organization of the elements that Labov and Waletzy (1967) and Labov (1972) do not exemplify. Labov (1963: 363) acknowledged the existence of "complex chainings and embeddings" of the narrative elements he identified; however, as he himself pointed out, he was "dealing with the simpler forms" of narratives.

Parsing narratives of discovery into the narrative elements outlined in Labov (1972) reveals that these stories do not always follow the order he observed in the narratives of personal experience; moreover, the structural elements in narratives of discovery demonstrate an organizing mechanism that includes sections of narratives similar to Labov's (1972) elements but that is more complex due to multiple interruptions between the elements.

The extension of the secondary structure with the addition of the Explanation doubles the opportunities for suspending the temporal sequence. The Explanation and Evaluation can interrupt a narrative of discovery almost at any point. Thus the secondary structure of narratives of discovery has multiple possible locations within a narrative and does not necessarily follow the complicating action as it does in narratives of personal experience (Labov 1972: 369–370).

This is not unexpected since popular science narratives of discovery are quite different from oral narratives of personal experience. Other researchers (see, for example, Peterson and McCabe 1983) also noticed deviations from Labov's order and even exclusions of some of the elements when narratives other than those of personal experience of adults were analyzed. Unlike Peterson and McCabe (1983), who analyzed personal experience narratives of small children, I did not observe narratives of discovery that lacked resolutions or proceeded from event to event without apparent connections. What I observed were narratives with the interrupted Complications, Orientations, and Results. In the example from Kaku (2011: 148) "Resveratrol," the Complication is interrupted by the Explanation. Notice also that the narrative ends with a cluster of the elements representing the scaffolding secondary structure. The secondary structure of this narrative makes this story's organization more complex, but at the same time it makes the narrative easier to understand for a non-specialist. It also gives the story meaning by telling the reader exactly why the discovery is important and what its outcome might be.

"Resveratrol"

1. Orientation
   What motivates scientists is the search for a gene that controls this mechanism, whereby we can reap the benefits of caloric restriction without the downside.

2.a. Complication
   An important clue to this was found in 1991 by MIT researcher Leonard P. Guarente and others, who were looking for a gene that might lengthen the life span of yeast cells. Guarente, David Sinclair of Harvard, and coworkers discovered the gene SIR2, which is involved in bringing on the effects of caloric restriction. This gene is responsible for detecting the energy reserves of a cell.

3.a. Explanation
   When the energy reserves are low, as during a famine, the gene is activated. This is precisely what you might expect in a gene that controls the effects of caloric restriction.

2.b. Complication
   They also found that the SIR2 gene has a counterpart in mice and in people, called the SIRT genes, which produce proteins called sirtuins.

4. Result
They then looked for chemicals that activate the sirtuins, and found the chemical resveratrol.

5. a. Evaluation
This was intriguing, because scientists also believe that resveratrol may be responsible for the benefits of red wine and may explain the "French paradox."

3.b. Explanation
French cooking is famous for its rich sauces, which are high in fats and oils, yet the French seem to have a normal life span. Perhaps this mystery can be explained because the French consume so much red wine, which contains resveratrol.

5.b. Evaluation
Scientists have found that sirtuin activators can protect mice from an impressive variety of diseases, including lung and colon cancer, melanoma, lymphoma, type 2 diabetes, cardiovascular disease, and Alzheimer's disease, according to Sinclair. If even a fraction of these diseases can be treated in humans via sirtuins, it would revolutionize all medicine.

In general, the Explanation is a useful element that is most pronounced in the narratives of discovery. At the same time, if you go back to chapter 2 now and look at the personal narratives with the new knowledge about the Explanation in mind, you will spot the carefully planted Explanations in Greene's and Coen's narratives. These are some of the most "technical" personal narratives; they have not only personal observations and experiences but also science, and almost every time science figures into a story, it requires an explanation. After all, this is why we read popular science—to have complex (and not so complex) scientific matters explained to us. It is easy to see why some scholars propose to use "explanation" as a blanket term for popular science writing. However, popular science has functions other than explanatory (see the following chapter), and explanations are reserved for specific sections of certain narratives.

Narrative technique is indispensable for communication of science to the public. Linguistic insight into the structure of popular science narratives, however, remains somewhat overlooked. Those who investigate popular science from the point of view of linguistics (see, for example, Moirand 2003; Myers 2003; Turney 2004a, b; De Oliveira and Pagano 2006; Fu and Hyland 2014) tend to address either broad issues such as explanatory properties (see, for example, Turney 2004b) or the general structure and effectiveness of a message (see, for example, Moirand 2003; Myers 2003). Others take a very narrow approach that addresses one specific linguistic issue which usually deals more with vocabulary choices or organization on a sentence level than it does with overall structure (see, for example, De Oliveira and Pagano 2006 or Urbanova 2012 for analyses of discourse presentation; Fu and Hyland 2014 for exploration of interactional metadiscourse).

General narratology usually regards scientific and popular scientific discourses as a side note (see, for example, Herman 2009).

It might be tempting, in the circumstances, to introduce an entirely new system that will focus on narrative structure in popular science. At the same time, as the analysis undertaken in this chapter shows, Labov's (1972) framework for the exploration of narrative macrostructure can be used productively to explicate popular science narratives. Not only that, it brings to the forefront the element of Explanation, an addition to the original model. The application of Labov's (1972) model is advantageous because it positions popular science narratives as parts of narrative tradition that goes back to the essential human need for storytelling.

Labov's (1972) model is not overly complex, yet it carries weight in academic circles to this day. Unlike many suggestions on narrative organization proposed by non-linguists (see, for example, Olson 2015), Labov's framework is based on solid scholarly research that has been confirmed over and over again by generations of language scholars. It is not a fad and does not fit into a catchy acronym, and, yes, it was not developed with popular science narratives in mind. However, in its distance from popular science and, therefore, in its neutrality, lies its strength. If a science writer uses the elements of Labov's model to shape her story, she will be creating not a narrative reflecting the latest trend in approaching her writing task but producing a text that will withstand the test of time and be recognizable as a properly structured narrative for years to come. In other words, the model is a classic.

All of that being said, Labov's (1972) framework is not without its limitations. In chapter 2, during the discussion of narratives of personal experience, I introduced the idea that Labov's (1972) and Hoey's (2001) models can be used together to classify complicating actions that occur in popular science narratives. For instance, the complicating action for the narrative "Galileo and Pendulum" (du Sautoy 2011: 26) contains the phrase "Galileo decided," and the complicating action for the "Duality" narrative (Greene 2011: 109–111) is described by "researchers could do nothing more than shrug, throw up their hands, and admit that the math they were using was too feeble to provide any reliable insight." Planning an action is one type of Complication, whereas admitting a lack of knowledge is significantly different.

Labov's (1972) model can be used to demonstrate differences in the complicating actions of popular science narratives, but it stops there. It does not supply the tools necessary to explain these variations nor does it provide analytical support for further investigation. As Ageliki Nicolopoulou

(1997: 375) notes, the Labovian analytical model has "the lack of a genuine interpretive dimension in its theoretical starting point." Nicolopoulou (1997: 369) suggests that as a result, "the tendencies in ... research that have drawn on Labov's analysis of narrative have done so in largely formalist and decontextualized ways." Tappan (1997: 380) also points out the limitations of Labov's (1972) framework when it comes to an analysis that attempts to reveal the meaning in addition to the narrative structure. Labov (1997) introduces several updates to the original model of narrative analysis; however, they too are primarily concerned with narrative structure and do not provide enough analytical tools for the full investigation of the narrative features the model helps to pinpoint. Hoey's (2001) patterns of narrative organization are much more suitable for that; moreover, they help expose an ideological side of popular science, and that is the subject of the next chapter.

4

Narratives and Ideology
What's in a Structure?

Popular science uses stories to shape your thinking about the scientific community. The authors choose narratives as vehicles for their ideological messages because, as Eggins and Slade (1997: 229) show, narratives are structured in a way that reveals "values, attitudes, and ways of seeing the world." We saw examples of this in the previous chapter with the use of the Evaluation element in narratives of discovery.

The presence of ideology in science, in itself, is not a novel idea. For example, Myers (1997) provides a clear demonstration of how a forum in which an account of a discovery is delivered can influence the readers' view of science. He shows that science can be presented as a gendered activity, where "women observe, while men experiment" (Myers 1997: 46). Science can also be described as a politicized activity (Myers 1997: 53). At the same time, science can be sold to the public as entertainment (Myers 1997: 50, 54).

Theatricality of science as a means of popularization and propagation of the idea that science can achieve amazing things that were once associated with magic was established in the 18th century and fully embraced in the 19th. At that time, as Coppola (2016: 9) notes, science began to be "offered as entertainment." Public lectures by Henry Pepper and Humphry Davy are great examples. Pepper used to present to his audience "a ghost"—a projection he created using optical apparatus. Davy, being a chemist, would create memorable explosions for the attendees of his lectures. Such remarkable visual displays were to illustrate, among other things, the omnipotence of science and to leave those who attended these events with a sense of awe. When written texts were produced based on the lectures, the scientists were careful to preserve the visual rhetoric of the experimental procedures.

Today, science does not strive to appear theatrical, at least not in print, but the ideology of omnipotence remains strong. Moreover, modern authors prefer to present scientific discoveries as proceeding via a planned series of steps leading to a predetermined goal, rather than as a serendipitous or an organic process. Showing scientists in full control of their environment assures positive results. This is what popularizations are after—convincing you that, for the most part, all scientific experiments produce desired results and that these results are always meaningful and propel the human race to further technological and scientific advances. In the chase for this kind of straightforwardness, scientific discovery gets boiled down to a process that includes an idea, collection of empirical data, and proof using the experimental results. When one of the steps in the process fails, so does the discovery. Only a small group of narratives showcases discoveries that happen not through this type of directed process but because of serendipitous events. Another small group of narratives reveals the role of human participants and their emotions towards each other and the scientific tasks they undertake.

These narratives showcase science's human side—or as Gilbert and Mulkay (1984) term it—the contingent repertoire of scientific discourse. At the same time, these narratives, too, conform to the overall message of science being a directed process.

To see how narratives are manipulated in order to promote the view of science as a directed process with an inevitably positive result, we need to turn to Hoey's (2001) patterns of structural organization that I introduced in chapter 2. Table 4.1 shows the three patterns most often found in popular science and gives examples of their lexical signals.

Table 4.1. Hoey's Text Patterns in the Narratives of Discovery

Type of Pattern	Examples of Lexical Signals
Problem-Solution	to baffle, to leave in shambles, to be unable, to drive crazy, to delay, to stump, could do nothing more, problem, hurdle, challenge, mistake
Gap in Knowledge-Filling	not to know, to become gripped by, to ask oneself, to ask, to lead to take a fresh look, to take on the challenge to determine, to be on the lookout for, to think, to establish whether
Goal-Achievement	to decide, to turn attention to, to anticipate, to speculate, to conjecture, to prove, to pursue, to envision, to have high hopes, goal

Popular science narratives of discovery may also combine two of these patterns to form a Gap/Goal narrative, for example, where the resolution would both fill the gap in knowledge and demonstrate the achievement of the goal. Such narratives use lexical signals that belong to both patterns. Another notable variation has to do with the Problem-Solution narratives and includes a focus on a specific type of problem—a conflict between scientists. Thus it is possible to talk about conflict narratives as a subtype of the Problem-Solution pattern. Hoey's (2001) patterns are also useful in singling out narratives with negative outcomes and explaining what the break down in the structure means. I label such stories narratives of failed discoveries.

Before I discuss the types of narratives mentioned above, I'd like to point out that it is possible to regard all of the various patterns as one. Since for scientists knowledge is a very important commodity or "social good," to use Gee's (2011: 5) terminology, the absence of that social good—gap in knowledge—is a very serious problem, and the scientists' goal is to achieve that knowledge. In this sense, the three patterns (Problem-Solution, Goal-Achievement, Gap in Knowledge-Filling) are actually one and the same when it comes to popular science narratives. The combining of the patterns further attests to the possibility of one overarching pattern. Hoey (2001: 152) calls it "pattern-complex."

Let's come back to Greene's (2011: 109–111) narrative about duality. It is a story of five different types of string theory which were believed to have no connections to one another; however, in 1995 Edward Witten "argued that when the coupling constant in any one formulation of string theory is dialed ever larger, the theory ... morphs into ... one of the other formulations of string theory, but with a coupling constant that's dialed ever smaller.... Which means that the five string theories are not different after all." This narrative begins with lexical signals that indicate Problem, Gap, and Goal simultaneously: "But outside of this restricted domain of small string couplings, researchers could do nothing more than shrug, throw up their hands, and admit that the math they were using was too feeble to provide any reliable insight."

At first, I classified this narrative as a Gap in knowledge-Filling narrative. At the same time, the fact that "the math ... was too feeble" is more than simply a gap in the scientists' knowledge; it presents a clear problem since it hinders research. The lexical signals are multi-functional and indicate both the gap in knowledge and a problem. The Result and Evaluation sections are combined in this discovery story and support its classification as a Problem-Solution narrative. In fact, the Result/Evaluation section restates what was previously presented as a gap in knowledge as a problem: "perturbative calculations in one string theory can't be undertaken because that theory's coupling is too large." The phrase "can't be undertaken" is much stronger than "could do nothing more than shrug, throw up their hands," which indicates lack of knowledge and not the actual inability. Considering the broader context of this narrative, it is possible to argue for the presence of the goal as well. After all, there are no indicators that the discovery happened by chance; quite the opposite, Greene (2011: 109) tells the reader that the discovery happed as a result of "drawing on the insights of scientists including Joe Polchinski, Michael Duff, Paul Townsend, Chris Hull, John Schwarz, Ashoke Sen, and many others." Thus Witten must have set a goal of solving the problem and finding the answer. Why else was he doing the research that lead to the discovery?

Hoey (2001: 140) acknowledges that all the patterns he describes "share many of its [the Problem-Solution pattern's] properties." In fact, Hoey (2001: 166) recognizes that "there may be circumstances in which distinguishing them [the patterns] is unnecessary." His suggestion is to call the ultimate pattern "SPRE," where S stands for situation, P for problem/gap/goal, R for result, and E for evaluation. Hoey (2001: 166), however, warns that analyzing texts in terms of this ultimate pattern "brings a greater degree of abstraction ... and takes us further from ... the detail of the lexical realization." This is true, but as he also acknowledges, "for some purposes [this] may be good." In the case of popular science narratives of discovery, regarding them as a broad category and looking at them in the context of science as a discipline makes clearer the point these narratives are making.

When the gap/goal/problem stories are considered as one category, a certain message emerges out of the narratives of discovery. It becomes clear that discoveries are presented as results of directed and carefully performed processes. In other words, everything goes according to plan, always. The lexical signals that are used to identify the patterns and their components (problem/gap/goal, solution, result) reveal the steps that scientists take in order to achieve breakthroughs.

At times, however, the authors opt not to follow the chronological order of the discovery process. Almost half of the narratives of discovery rely on a modification of the linear order of the steps. It is not uncommon for a narrative to present a summary of the pattern (for example, name the problem and immediately indicate that it was solved) and then continue to explain how the result was reached. This structure echoes what Hoey (1983:134) and McCarthy (2005: 158) call "General-Particular" and "general-specific" pattern. Hoey (1983: 138) suggests a subtype of this pattern that helps explain the organization of many of the narratives of discovery. He calls the subtype "The Preview-Detail" pattern. Consider the following narrative:

"String Theory and Black Holes"

1. Introduction
Nevertheless, one major advance has illuminated a related aspect of black holes ... the work of Jacob Bekenstein and Stephen Hawking in the 1970s established that black holes contain a very particular quantity of disorder, technically known as entropy. According to basic physics, much as the disorder within a sock drawer reflects the many possible haphazard rearrangements of its contents, the disorder of a black hole reflects the many possible haphazard rearrangements of the black hole's innards.

2. Problem
But try as they might, physicists were unable to understand black holes well enough to identify their innards, let alone analyze the possible ways they could be rearranged.

3. Solution—general (preview)
The string theorists Andrew Strominger and Cumrun Vafa broke through the impasse.

4. Solution—specific (detail)
Using a melange of string theory's fundamental ingredients..., they created a mathematical model for a black hole's disorder, a model transparent enough to enable them to extract a numerical measure of the entropy.

5. Result
The result they found agreed spot-on with the Bekenstein-Hawking answer. While the work left open many deep issues (such as explicitly identifying a black hole's microscopic constituents), it provided the first firm quantum mechanical accounting of a black hole's disorder.

6. Evaluation
The remarkable advances in dealing with both singularities and black hole entropy give the community of physicists well-grounded confidence that in time the remaining challenges of black holes and the big bang will be conquered [Greene 2011: 98].

By the time the reader reaches section 3, he/she knows that the problem has been solved. Section 4 explains how the solution was reached, and sections 5 and 6 evaluate and reaffirm the solution as a positive outcome. This is not an unusual route for a popular science author to take. The Preview-Detail organizing principle can be observed in a narrative following any of the three patterns (Problem, Gap, Goal).

Supplying the reader with a summary or a preview of the narrative's outcome indicates an important angle that popular science narratives of discovery are trying to present. The point of these stories is not so much to intrigue the reader as to whether or not the problem will be solved but to promote an image of science and scientists as capable or resolving any difficult situation. By placing the outcome of the story in the beginning of the narrative, the authors dismiss the possibility that a problem/gap in knowledge/goal can be left completely unresolved. Even when the filling of the gap in knowledge is not complete, the partial knowledge is presented in a positive light. Consider an example from du Sautoy (2011:45): "Ever since the Greeks had proved that the primes go on forever, mathematicians had been on the lookout for clever formulas that might generate bigger and bigger primes. One of the best of these formulas was discovered by a French monk named Marin Mersenne.... Unfortunately for Mersenne and for mathematics, his idea didn't quite work.... But although Mersenne's idea didn't always work, it has led to some of the largest prime numbers that have been discovered."

Relying on the Preview-Detail pattern also means that the authors of popular science write stories that don't always follow the actual chronological order of the events that took place. This is both unusual and expected. According to a definition of a prototypical narrative: a story is a chronological sequence of events connected in a meaningful way. At the same time, how many stories can you think of that start at the end and then unravel into explanations of how this particular end came to be? A preview is a popular narrative device. In fact, it has been in the repertoire of popular science for a while. Curtis (1994) did a study suggesting that popular science narratives resemble detective stories. Detective stories are all about solving the murder, not wondering whether or not a murder will happen. The same is true for narratives of discovery. It is not always the anticipation of a breakthrough that drives these stories but the details of the process.

So what kind of process is it? You might guess that the authors would introduce descriptions of laboratory experiments: the instruments used, the activities performed. If so, you will be wrong. Discovery narratives tend to emphasize the following steps: an idea or insight, experimental results, and proof of the initial idea based on the results of the experiment. The first step (the idea) is usually set by the lexical signals of problem/gap/goal. The second step (experimental results) is demonstrated by the lexical signals that indicate what was done about the problem/gap/goal. And while step two presupposes a description of experimental procedures, it is usually the most underdeveloped and unrealistic portion of the story. The third step (proof) is signaled by the vocabulary that also indicates the resolution of the narrative. This process is clearly a result of the application of the scientific method. However, the actual experimental procedures and the apparatus associated with them rarely make it into the stories. The deliberate focus on the ideas and results is not accidental.

First of all, by not emphasizing the laboratory environment and the physical actions associated with experiments, the authors are placing greater importance on the intellectual actions than on the work that confirms them. Second of all, this is doubling down on the message about the power of science. The scientific community has all the answers; how these answers materialize is not that important. As a result, the readers face a greater number of narratives of discovery that foreground the intellectual actions of scientists rather than experimental procedures. Out of the narratives that describe science as a directed process (there is a small group that breaks this mold), 63 percent emphasize the importance of an idea or an intellectual activity over the experimental step. In such narratives, the experimental part is usually reduced to a statement that proof was obtained. Compare the description of Newton's discovery of gravity to that of Rice University's scientists' discovery of how to make nanotubes:

> He [Newton] realized that the forces that grab an apple are the same that reach out to the planets and comets in the heavens. This allowed him to apply the new mathematics he had just invented, the calculus, to plot the trajectory of the planets and moons, and for the first time to decode the motions of the heavens [Kaku 2011: 298]
>
> They discovered, by trial and error, that these carbon nanotubes can be dissolved in a solution of chlorosulphonic acid, and then shot out of a nozzle, similar to a shower head. This method can produce carbon nanotube fibers that are 50 micrometers thick and hundreds of meters long [Kaku 2011: 280].

The pattern signals for the narratives presenting discoveries as intellectual activities will always be different from the pattern signals for stories that emphasize laboratory procedures. To return to our examples, the "Newton and Gravity" is a gap-filling narrative with the following signal that triggers the pattern, "asked himself a question"—a purely intellectual action. The signals for the filling of the gap (the outcome of the story) are also intellectual actions, "he realized," "allowed ... to decode," "introduced a new way of thinking." Upon finishing this narrative, the reader does not know how exactly Newton came to his conclusions, all he or she knows is that the famous scientist "realized that the forces that grab an apple are the same that reach out to the planets and comets in the heavens." The experimental step is a mere mention of mathematical analysis, "This allowed him to apply the new mathematics he had just invented, the calculus, to plot the trajectory of the planets and moons." The reader is spared the actual details of the analysis, and the final filling of the gap signal "to decode" reinforces the focus on the intellectual processes in this narrative: the verb is used metaphorically.

The "Composite Nanotubes" narrative works differently. The very pattern trigger, "the problem is creating," suggests physical actions. The signals of the solution to the problem are experimental actions "dissolved" and "shot out." These are not intellectual processes, and the narrative contains a somewhat detailed description of the experiment the scientists used to solve the problem stated at the beginning of the story.

The dominance of narratives with intellectual actions (like the one about Newton) might be explained by the subjects of certain stories. For example, a narrative about string theory is bound to focus on intellectual processes because little experimental proof exists yet. Many mathematical problems are solved not through physical experimental work but by solving equations, so narratives describing such breakthroughs will, also, inevitably emphasize intellectual actions. However, in biology, for example, it is difficult to make a discovery without conducting any physical experiments. In this case, one expects a story to include a description of the experiment and maybe even of the apparatus used. However, much more often, even in cases where the discovery was achieved by the means of an experiment, the details of physical manipulation of equipment are omitted in favor of describing the intellectual activities behind them. For example, in the story about Francis Crick, James Watson, Donald Caspar, and Aaron Klug discovering the shapes of viruses, du Sautoy (2011: 69) implies that an experimental procedure led to the discovery but never describes it. The narrative makes it clear that the discovery was made with the help of certain tools. The mentions of "the two-dimensional pictures from X-ray diffraction," "images," and "another diffraction image" suggest that the discovery was not purely an intellectual activity and that the scientists obtained their data as a result of an experiment. Moreover, this experiment required manipulation of some kind of scientific apparatus, which is never mentioned. However, it appears that the conclusions scientists draw are the results of intellectual analytical processes alone. To reinforce this idea, the actions of the scientists in this narrative are mostly intellectual (bolded). Even when the verbs describing them denote physical activities such as "cracked," "looked," "found," "home in," they are used in their metaphorical senses to describe intellectual activities:

> Having **cracked** the structure of DNA, Francis Crick and James Watson, along with Donald Caspar and Aaron Klug, turned their attention to what the two-dimensional pictures from X-ray diffraction could reveal about viruses.
> When the biologists **looked in the mathematicians' cabinet of shapes**, it was the Platonic solids that seemed the best candidates for the form of these viruses.
> To their surprise, they **found** shapes full of symmetry.
> Suddenly, dots arranged in pentagons appeared, and that **allowed them to home in** on one of the more interesting of Plato's dice: the icosahedron—the shape made of 20 triangles with five triangles meeting at each point.

Besides relying on metaphorical use of physical action verbs, narratives of intellectual processes have another peculiarity. Not surprisingly, in such stories an experiment is never the central part of the discovery. What's more, sometimes, the focus on thinking rather than on doing produces stories where intellectual/theoretical confirmation is presented as more important than experimental proof. For example, Greene (2011: 48–49) includes the following detail in his narrative about magnetic fields:

> Unexpectedly, Faraday's experiments showed that electric and magnetic fields are intimately related: he found that a changing electric field generates a magnetic field, and vice versa. In the late 1800s, James Clerk Maxwell put mathematical might behind these insights, describing electric and magnetic fields in terms of numbers assigned to each point in space; the numbers' values reflect the field's ability, at that location, to exert influence. Places in space where the magnetic field's numerical values are large, for instance an MRI cavity, are places where metal objects will feel a strong push or pull. Places in space where the electric field's numerical values are large, for instance the inside of a thundercloud, are places where powerful electrical discharges such a lightning may occur.

This is, essentially, an example of the reversal of the steps in the process of discovery. In this case, the experiment came first, "Faraday's experiments showed that electric and magnetic fields are intimately related," and the theoretical explanation (the understanding or the grand idea) came after, "James Clerk Maxwell put mathematical might behind these insights." The experiment is referred to as "insights," and the theoretical confirmation is described as "might." Such structure and wording imply that the discovery was not fully valid until the theoretical confirmation and almost suggests that Faraday, the experimenting scientist who discovered electromagnetism, didn't quite know what he was doing. His contribution is a side note in the story of Maxwell's mathematical analysis.

Such explicit focus on intellectual actions, no matter the subject of a narrative, and the intentional de-emphasizing of the physical experimental work shows that the authors of popular science present the first step of the discovery process (the idea—the intellectual action) as the more crucial to the success of the whole process than the subsequent steps. That is not to say that narratives describing experimental work are nonexistent—but their number is lower than that of the narratives showcasing intellectual actions: 37 percent vs. 63 percent. These percentages, taken from a relatively small sample, do not indicate a definite conclusion but rather show a trend: popular science books promote the view of science as primarily an intellectual activity, which is supplemented with experimental work. This is not a phenomenon reserved entirely for popular science books or found exclusively in narratives, for that matter. Popular science blogs and articles also lean this way, glorifying analysis and undermining experimental work.

It is possible to attribute the volume of the narratives emphasizing analytical skills to the possibility that the authors find such stories the most interesting. As Polanyi (1979: 211–212) suggests, "interestedness" is culturally, socially, and personally determined. Thus the dominance of the intellectual activity-centered narratives might be an indicator of what the authors consider interesting based on their professional backgrounds. It is possible to interpret these findings as reflective of the authors' opinions of what is most appealing about scientific discoveries.

Whether narratives of discovery emphasize intellectual actions or physical experiments, the focus on the process remains. The only difference is literally in the details of experimental work—they could be described explicitly or mentioned briefly, but the step that requires the obtaining of proof is present in all the process narratives. The importance of the message that discoveries are achieved by following specific steps becomes especially clear when the narratives of failed discoveries are considered.

Narratives of failed discoveries show what happens when the process of scientific discovery fails. In such narratives, either the ideas do not find experimental support, or the experimental results are not recognized as significant. While there aren't many stories like this, they are worth the mention because they place the greatest value on the combination of experimental work and the ideas behind it. Structurally, they follow the SPRE (Situation, Problem, Result, Evaluation) pattern. The narratives of failed discoveries serve as an example of why it might be useful to regard the problem/gap/goal pattern as a single category since the lexical signals in these narratives are not specific enough to categorize the stories as following either Gap in Knowledge-Filling, Goal-Achievement, or Problem-Solution patterns.

Below is a story of Ronald Richter, a scientist who claimed to have discovered cold fusion:

"Argentinean Fusion"

Situation
  Back in 1951, when the United States and the Soviet Union were gripped in Cold War frenzy and were feverishly developing the first hydrogen bomb,

Problem
  President Juan Peron of Argentina announced, with huge fanfare and a media blitz, that his country's scientists had made a breakthrough in controlling the power of the sun. The story sparked a firestorm of publicity. It seemed unbelievable, yet it made the front page of the *New York Times*. Argentina, boasted Peron, had scored a major scientific breakthrough where the superpowers had failed. An unknown German-speaking

scientist, Ronald Richter, had convinced Peron to fund his "thermotron," which promised unlimited energy and eternal glory for Argentina.

Evaluation

The American scientific community, which was still grappling with fusion in the fierce race with Russia to produce the H-bomb, declared that the claim was nonsense. Atomic scientist Ralph Lapp said, "I know what the other material is that the Argentines are using. It's baloney." The press quickly dubbed it the Baloney Bomb. Atomic scientist David Lilienthal was asked if there was the "slightest chance" the Argentines could be correct. He shot back, "Less than that." Under intense pressure, Peron simply dug in his heels, hinting that the superpowers were jealous that Argentina had scooped them.

Result

The moment of truth finally came the next year, when Peron's representatives visited Richter's lab. Under fire, Richter was acting increasingly erratic and bizarre. When inspectors arrived, he blew the laboratory door off using tanks of oxygen and then scribbled on a piece of paper the words "atomic energy." He ordered gunpowder to be injected into the reactor. The verdict was that he was probably insane. When inspectors placed a piece of radium next to Richter's "radiation counters," nothing happened, so clearly his equipment was fraudulent. Richter was later arrested [Kaku 2011: 236].

The Evaluation and the Result invalidate the claims of discovery made earlier and turn what could have been a standard discovery story into a narrative of failed discovery. And while a negative Result is typical for such narratives, not all of them start out by doubting the main idea as this narrative does: Kaku (2011: 236) reflects the attitude of the scientific community at the time: "It seemed unbelievable."

Narratives of failed discovery are excellent cases to test Hoey's (2001: 130–133) proposition that when a story reaches a negative outcome and/or evaluation, "the pattern has to recycle" until the initial problem is resolved in a positive way. At first it appears that popular science narratives of discovery do not fully follow this scenario. The only example that resembles pattern recycling among the narratives I analyzed is du Sautoy's (2011: 23) tale about zero. The story first reaches a negative outcome, "The Babylonians introduced a little symbol" that functioned somewhat like a zero, "but they didn't think of 0 as a number in its own right. For them, it was just a symbol used in the place-value system to denote the absence of certain powers of 60." Then, without going through the pattern steps again, the author introduces a positive resolution, "Mathematics would have to wait another 2,700 years until the seventh century AD, when the Indians introduced and investigated the properties of 0 as a number."

du Sautoy's (2011: 23) is not a typical narrative of failed discovery. In his story, the discovery fails not because there was no experimental support for the initial idea but because the discoverers failed to recognize the importance of their achievement and develop it further. The break down of the process occurred not at the experimental but at the conceptual stage.

Another observation that Hoey (2001: 132–133) makes about stories with negative outcomes is that in some cases the pattern can end even if the reached outcome is unsatisfactory, but only if the outcome is "beyond retrieval." Greene's (2011: 73) narrative of Einstein's quest for the unified theory is a good example of such a case. Greene tells his readers that "during the last thirty years of his [Einstein's] life the problem of unification became his prime obsession." However, even Einstein, a man who had an "unparalleled capacity for single-minded devotion to problems he'd set for himself," could not solve that particular problem. The famous scientist spent the last hours of his life working on the unified theory, yet "his final scribblings shed no further light on unification." There was no discovery. Yet the narrative is complete, and so is the pattern even though the outcome is negative. Since the main character died, there is nothing more he can do about reaching his goal; the negative outcome is "beyond retrieval," and no one else has solved the problem conclusively since. This narrative that does not provide a positive outcome and does not recycle the pattern, not even partially, still conforms to the overall theme of failed discovery narratives. It shows the break down in the process of discovery. Einstein had an idea, but he failed to support it with mathematical proof.

Narratives of failed discoveries may be few, but they play an important role and support the idea of science as a process. These stories show what happens when there is a breakdown in one of the links of the process chain—be it the failure to find empirical confirmation or the failure to recognize the importance of experimental results. Another noteworthy point about these stories is that narratives of failed discoveries do not stress one part of the process and undermine another; they emphasize the importance of both the intellectual and the experimental parts of doing science.

One more variation among the process narratives is represented by the discovery narratives that demonstrate conflict among the members of the scientific community. These narratives, while showing scientific discoveries as directed processes, also present science as a human activity, more so than other types of stories do.

It has been established that popular science texts tend to emphasize human participants and that scientists are often presented as "personalities" (Parkinson and Adendorff 2004: 388–389). Narratives of discovery that include conflict focus on the scientists themselves the most; however, these narratives include not interpersonal conflicts but scientific disagreements. They demonstrate the scientists' desire for their ideas to be recognized, thus establishing recognition as an important social good within the scientific community. Denial of this social good often causes a conflict, and the resolution of this conflict involves accepting the previously ignored claims.

Conflict narratives could be regarded as Problem-Solution narratives, and they continue the theme of science as a process that begins with an idea which is later supported by experimental results. The major difference between Problem-Solution narratives and conflict narratives is the nature of the problem and its outcomes. In a prototypical Problem-Solution narrative, the problem and its resolution lead to a discovery; in a conflict narrative, the problem is not a precursor of a discovery but a consequence. A conflict, usually arises when a discoverer is ignored or disbelieved, or when a discovery contradicts previous assumptions about a phenomenon, which no one but the discoverer is willing to give up. Let's look at an example—Greene's (2011: 189–191) story about Hugh Everett III, "an unknown graduate student from Princeton University," whose PhD dissertation "was one of the earliest mathematically motivated insights suggesting that we might be part of a multiverse."

As is typical for conflict narratives, the discovery is introduced at the beginning. It will set off both the scientific conflict and a personal problem. When Everett submitted his discovery and his thesis to "John Wheeler, his doctoral adviser[,] Wheeler, one of twentieth-century physics' most celebrated thinkers, was thoroughly impressed." In fact, he admired Everett's ideas so much that he decided to share his star-student's achievements with quantum physics' "luminaries like Niels Bohr."

This is where the scientific conflict begins: "the reception was icy. Bohr and his followers had spent decades refining their view of quantum mechanics. To them, the questions Everett raised, and the outlandish ways in which he thought they should be addressed, were of little merit." The lexical signals in this portion of the story clearly signal conflict and are expressed by adjectives, "icy," "outlandish," "of little merit." They indicate that the scientific community rejected Everett's discovery.

This, in turn, prompted a personal problem for Everett: "In response to the criticisms, Wheeler delayed granting Everett his Ph.D." The lexical signals for the problem are expressed differently than those for conflict. We now see action verbs as the signaling vocabulary. As the doctoral adviser "compelled him to modify the thesis substantially," Everett at first "resisted," but then "reluctantly acquiesced." The lexical signals of problem—"compelled," "resisted," and "acquiesced"—trace the trajectory of a sub portion of this story which deals with Everett getting his Ph.D. The personal

4. Narratives and Ideology 83

problem is resolved when, "in March of 1957, Everett submitted a substantially trimmed-down version of his original thesis; by April it was accepted by Princeton as fulfilling his remaining requirements." At the same time, the scientific conflict remained.

Everett, despite negative comments from Bohr, published his research "in the *Reviews of Modern Physics*." The paper that appeared, however, was not "the grander vision articulated in the original thesis" but a distilled outline of the multiverse proposal. This and the fact that "Everett's approach to quantum theory ... [had] already been dismissed by Bohr and his entourage," meant only one thing—"the paper was ignored."

The conflict with the scientific community remained unresolved since Everett did not receive recognition for his discovery.

This story does not end here. As Hoey (2001) suggests, when a narrative reaches an unsatisfactory outcome, the pattern has to recycle. And luckily for Everett, his story goes on. Ultimately, in conflict narratives the conflict cannot be resolved by the main character alone since he/she requires that the other members of the scientific community accept the discovery. Sometimes the scientists who are involved in the resolution of the conflict are mentioned by name, and at other times they are simply referred to as "the physics community," for example (Greene 2011: 41).

The second time around, Everett's story also involves a graduate student—Neill Graham—and his adviser—"the renowned physicist Bryce DeWitt." Graham helped with the proof for the multiverse idea as he "further developed Everett's mathematics," and DeWitt organized the right kind of publicity for the discovery—he involved the public. "Besides publishing a number of technical papers that brought Everett's insights to a small but influential community of specialists, in 1970 DeWitt wrote a general level summary for *Physics Today* that reached a much broader scientific audience." In these publications, the now-popularizer "underscored.... Everett's conclusion that we are part of an enormous 'multiworld.'" In the end, it was the article targeting a broad audience that brought recognition to Everett's work: "The article generated a significant response in a physics community." This story is not only a tale with a scientific happy ending; it is also a confirmation of the power popular scientific outlets hold, which I discuss in the introduction.

In analyzing conflict narratives, it is possible to assume that they are focusing on the emotions and the personalities of the scientists and thus stand in opposition to the process narratives. However, when the texts are considered carefully, it becomes clear that conflict narratives focus on the intellectual ideas much more than they do on the scientists as individuals.

To continue with the "Many Worlds" example from Greene (2011: 189–191), the full name of the main character (Hugh Everett III) is mentioned once, and from then on, the name or the references to it appear in the possessive case more often than they do in the nominal (11 instances of nominal case, 17 instances of possessive mention). This indicates that the narrative makes more references to the ideas and contributions of the scientist than it does to the scientist himself. It is especially noteworthy that the last section of the narrative that includes the resolution of the conflict does not have any nominal references at all. In this narrative, Everett is primarily the source of ideas, not a participant in the story. The victory achieved at the end of the story is not personal; it is a vindication of the idea.

The overall focus of conflict narratives on scientific ideas shows up in the cause of the conflicts as well. Interpersonal problems are always grounded in scientific objections. That is the case in "Many Worlds," where "*the questions* Everett raised, and the outlandish ways in which he thought they should be addressed, were of little merit." The conflict is not with the scientist himself but with his ideas. This is made even clearer in the story of a disagreement between a priest and a scientist.

When Greene (2011: 11–12) introduces Georges Lemaître's discovery that "the universe began as a tiny speck of astounding density … which swelled over the vastness of time to become the observable cosmos," he makes it into a narrative of conflict between Lemaître and Einstein. The author provides an evaluation of Lemaître as an "unusual figure" and includes details of his theological background: "By 1923, he had not only completed his work for a doctorate, but he'd also finished his studies at the Saint-Rombaut seminary and been ordained a Jesuit priest." However, the conflict arises not over Lemaître's religion but over his scientific ideas: "Your mathematics is correct, but your physics is abominable," Einstein tells the man with a "clerical collar in place."

Later on, Einstein is not convinced by Lemaître's proof, and the famous scientist, not the priest, is the one who decides to follow "his deepseated belief that the universe was eternal and, on the largest of scales, fixed and unchanging." While Lemaître is dealing with a mathematical proof, Einstein "balked at those solutions" and "bucked the equations in favor of his intuition about how the cosmos should be." The ultimate conflict signal is Einstein's reprimand, "The universe, Einstein admonished Lemaître, is not now expanding and never was."

The resolution of this story deals with both Einstein's recognition of Lemaître's contribution to cosmology and the scientific community's acceptance of the discovery. Thus, the emphasis of the narrative is not solely or primarily on an a disagreement between two scientists, but it is the general approval of Lemaître's ideas by the scientific community that is at stake. At the end of the narrative, Einstein voices the importance of Lemaître's contribution: "Einstein stood up and declared Lemaître's theory to be 'the most beautiful and satisfactory explanation of creation to which I have ever listened.'" And there is also an indication that the whole scientific community embraced Lemaître's discovery: "While still largely unknown to the general public, Lemaître would come to be known among scientists as the father of the big bang."

While neither Everett's nor Lemaître's story describes a strictly personal conflict, there are overtones of interpersonal struggles in both narratives. The disagreement arises between a young and less experienced scientist and the more seasoned members of the scientific community. The more experienced scientists are more likely to make assumptions and judgments, while the younger researchers do not have the luxury to make arguments based on intuition or personal beliefs. If they want to be taken seriously, they have to have concrete proof on their side, and even then, as we see in the two narratives, it is not an easy road to recognition.

Everett is up against Bohr, who is described as "the grand master of quantum mechanics" and a luminary, while Everett is introduced as "an unknown graduate student from Princeton University." Lemaître, in addition to being a younger scientist, is identified as "a Jesuit priest," which puts him in opposition to Einstein on more than one level. These stories are an excellent illustration of science as a human activity where interpersonal relations play an important role, but where ultimately the best ideas triumph.

As a group, conflict narratives of discovery reveal the human interactions involved in scientific progress. However, they do not overemphasize the personalities of the scientists but treat the scientists as the sources of ideas. In this respect, they continue the theme of the importance of ideas in the process of scientific discovery. At the same time, the young vs. experienced scientist theme suggests that science is not devoid of human emotions and judgments.

Conflict narratives represent what Gilbert and Mulkay (1984: 57) call "contingent repertoire," an informal scientific discourse that is heavily dependent on "personal or social circumstances." While the whole genre of popular science falls within this particular subtype of scientific discourse, the narratives of discovery that exhibit its core elements (subjectivity, personal judgment, role of social position) more explicitly than any other group are the conflict narratives. According to Gilbert and Mulkay (1984: 57),

"When this [contingent] repertoire is employed, scientists' actions are ... depicted ... as the activities and judgements of specific individuals acting on the basis of their personal inclinations and particular social positions." This is what leads to conflict in the first place.

As Gilbert and Mulkay (1984: 56) observed, on the other end of the spectrum is the formal, or "empiricist discourse," which creates the image of an objective, "generic" response of a scientist to the phenomena of the natural world. The narratives of discovery following the Problem/Gap/Goal pattern and depicting scientific discoveries as a directed process (following the steps of the scientific method) are somewhat closer to the empiricist discourse since they do not emphasize the role of personal judgment or social position. The conflict narratives, on the other hand, are often organized around the elevated social position of one scientist and the low social rank of another.

By far the majority of the narratives support the idea of science as a directed process. They present discoveries as a result of theoretical work confirmed by experimental results. Some variations of these narratives (conflict narratives and narratives of failed discoveries) show how the process of discovery can be affected by human emotions and demonstrate what happens when one link in the process fails to produce the required results. Even though these narratives allow for slight alterations to the process, they still demonstrate the authors' faith in it since roughly 93 percent of the narratives analyzed fall within the process group.

It would be possible to say that the directed process is the only way to describe a scientific discovery if not for a small group of narratives devoted to discoveries that happen because of serendipity and chance. These narratives show that a discovery can occur unexpectedly, for instance, as a result of a simple question by a non-scientist as in the story of chaos theory told by du Sautoy (2011: 229–230).

Poincaré did not set out to come up with chaos theory. He was solving a problem set up by "King Oscar II of Sweden and Norway" in 1885. The king wanted to "establish mathematically once and for all whether the solar system would continue turning like clockwork, or whether it was possible that at some point, the earth might spiral away from the sun and off into space." When Poincaré's proof was ready to be published, "one of the editors couldn't follow Poincaré's mathematics and raised a question." That single inquiry by a layman made the scientist identify an egregious error in his calculations. As du Sautoy writes, "It all seemed a huge embarrassment. But as often happens in mathematics, when something goes wrong, the reason it goes wrong leads to interesting discoveries…. What he discovered through his mistake led to one of the most important mathematical concepts of the last century: chaos theory."

This condensed version of the narrative du Sautoy presents to his reader clearly illustrates the role of chance in scientific discoveries. However, my retelling omits an important feature of such narratives. Pattern analysis of chance narratives shows that the serendipitous discoveries have a double structure. The first half of the narratives typically follows the Problem-Solution pattern. The second half of the narrative is usually triggered by a new problem that materializes as a result of the solution proposed in the first half of the story. In other words, there is pattern recycling at work. My summary introduced only the very beginning and the very end of the narrative: the bookends. It left out Poincaré's initial problem. Here it is: "Poincaré began by considering a system with just two bodies.... The problem is that as soon as you have three bodies in a system, for example the earth, moon, and sun, the question of whether their orbits are stable gets very complicated—so much so that it had stumped even the great Newton. The problem is that now there are 18 different ingredients to combine in the recipe: the exact coordinates of each body in each of three dimensions, and their velocities in each dimension." In other words, the calculations required to offer the kind of solid proof that the king requested were extremely complex. And, of course, Poincaré made an error, plus even with erroneous calculations "he couldn't solve the problem in its entirety." The partial resolution to the initial problem, however, receives a positive evaluation: "ideas were sophisticated enough to win him King Oscar's prize." Thus the recycling of the pattern is not prompted by the unresolved initial problem but by the interference of another character—the question by an editor, "Could Poincaré justify why making a small change in the positions of the planets would result in only a small change in their predicted orbits?" This very interference is the serendipitous event that triggered the discovery.

It is quite common for a problem, or rather its solution, to trigger a chance discovery. However, it is important to distinguish these narratives from the prototypical Problem-Solution stories that describe directed processes. In the narratives of chance discoveries, the problem that ends up leading to a discovery occurs unexpectedly, and the main character does not start out with a goal of solving this particular problem. The discovery of penicillin, as usually told (see, for example, Avraamidou and Osborn 2008), is a classic example. In fact, in some such narratives, the initial goal is only mentioned and not described in any significant detail. For instance, in the story of how Clinton Davisson and Lester Germer discovered the wave-like properties of electrons, Greene (2011: 193) does not dwell on the details of the initial experiment. All he says is that the discovery happened "in April 1925 during an experiment at Bell Labs" where "Davisson and Germer had been spending their days firing beams of electrons at specimens of nickel to investigate various aspects of the metal's atomic properties." This description of the experimental procedure sheds no light on the overall research goal for which the experiment was designed. Why did the scientists need to explore nickel's atomic properties? What was the end goal of their research? We never find out. Since it is not important to the narrative, the original research goal is unstated in favor of the problem that initiates a series of "fortuitous" events that result in the discovery. "A glass tube containing a hot chunk of nickel suddenly exploded," and upon cleaning the sample of nickel (they had to "vaporize the contaminant"), the scientists continued with their efforts. "But that choice, to clean the sample instead of opting for a new one, proved fortuitous. When they directed the electron beam at the newly cleaned nickel, the results were completely different from any they or anyone else had ever encountered." Davisson and Germer discovered that electrons can behave like waves. "The idea is," Greene (2011: 197) explains, "that in analyzing the motion of a particle we ... should think of it as a wave undulating from here to there."

Serendipitous narratives with this structure possess an element of surprise for the reader because the discovery is not revealed until the final step in the pattern—it is presented in the Result. But this is not so unusual for any type of a discovery narrative since the preview-detail pattern, where the discovery is revealed first and the route to it explained later, is not universal. While chance narratives do not necessarily describe a directed process like the majority of the narratives do, they still share structural similarities with the larger group.

The small percentage of serendipitous narratives no doubt reflects the scientific reality. How often does a discovery happen by chance? On the other hand, chance discoveries make for entertaining stories, and science—both popular and professional—likes to capitalize on good stories for publicity purposes. As Olson (2015: 9) puts it, "The journals want to tell good stories, the scientists want to tell good stories, ... and the journalists want to tell good stories." So including more narratives of serendipitous discoveries might be desirable. However, there would be a downside to such presentation of science, too. The fact that the authors of popular science books opt to focus on the process narratives shows their dedication to the presentation of science as a deliberate and controllable activity. They do not want the reader to think that scientific discoveries happen by chance. The

4. Narratives and Ideology   89

authors forgo the opportunity to dwell on the more intriguing tales in favor of promoting the image of a scientific community that is ultimately in control of the physical environment it investigates. Note that the triggers for chance discoveries are often undesirable events—a lack of understanding, apparatus malfunction. These are uncontrollable elements of everyday science that the authors are trying to purge from their narratives as they do not help maintain the "smiling face" presentation of science. It is not unusual for the authors to avoid such stories as much as possible.

On the other hand, it is possible to view serendipity in scientific discoveries as a positive force. Merton and Barber (2004) supply a detailed study of the role serendipity plays in science in general and in discoveries in particular. One of their findings suggests that serendipity and luck favor the prepared scientist. As Merton and Barber (2004: 171) explain, "to preparedness may be linked such qualities as alertness, flexibility, courage, and assiduity." A positive view of serendipity, Merton and Barber (2004: 84–186) argue, leads to a greater degree of coverage for experimental procedures, while a negative evaluation of serendipity results in an emphasis on ideas. If we follow this line of reasoning, it is possible to conclude that the narratives of discovery in popular science books have a negative outlook on serendipity. However, when only narratives of chance discoveries are considered, the data are inconclusive, as these narratives are equally divided between the ones that focus on experimentation and the ones that focus on ideas.

The analysis performed in this chapter goes one step beyond the investigation undertaken in chapter 3, where we focused exclusively on the structure of the discovery stories. In this chapter, we see that narrative structure can be used strategically to imbed a particular message. In the case of discovery narratives, the message is the omnipotence of science. Examined according to Hoey's (2001) story patterns, narratives of discovery in popular science suggest that the majority of discoveries happen as a result of a directed, deliberate, and fully controlled process initiated and carried out by scientists who set out to improve human understanding of the physical world around them. The only alternative to this road to discovery manifests in a small group of chance or serendipitous narratives. Such narratives show that discoveries can happen as a result of unexpected circumstances or equipment malfunction—a view that undermines the central message of popular science.

Narratives of discovery contain positive bias. It shows up as a deliberately celebratory presentation of science that excludes many of the difficulties and failures associated with doing research in the real world. Earlier studies (see, for example, Bucchi 1998, Dahl 2015, Fu and Hyland 2014, Koteyko et al. 2008, Moirand 2003, Turney 2004a) suggested the presence of positive bias on the sentence level, limited primarily to the words used, and as an activity associated with individual authors. In this chapter, I show that the choice to employ almost exclusively celebratory discourse goes deeper, and evidence of it could be found in the narrative structure itself. Positive bias is most easily observed through the use of preview-detail patterns and pattern recycling.

While structural analyses of the celebratory nature of popular and professional science exists (see, for example, Curtis 1994, Harré 1994), lately the attention has been focused primarily on the words that showcase the prowess of science. The ideological potential of narrative structure has been somewhat overlooked while at the same time the use of narrative technique has become a popular means of delivering scientific results not only in science's popular outlets but in professional publications as well (see, for example, Reitsma 2010, Blanchard et al. 2015, Hermwille 2016). This suggests that the high degree of objectivity automatically associated with scientific writing should, perhaps, be reevaluated. I discuss this issue in more detail in the conclusion.

5

What They Say
Speech of Scientists

In chapter 1, I introduced the basic modes of discourse presentation. Now has come the time to look at each one of them (speech, thought, and writing) in detail. In this chapter we will explore presented speech of scientists. Typically, using voices of scientists instead of narration adds authenticity and authority to a popular science text. I will show that presentation of speech can also be used in popular science to dramatize events and will suggest that in some cases such dramatization amounts to infusing a text with a healthy dose of fiction. Thus besides adding gravitas to popular science books, presented voices also help shape scientists into characters that the readers can relate to. Before we begin, let's consider how presented discourse is treated in fiction and non-fiction.

Discourse presentation in fiction is used to help the reader co-experience the events the characters are living through. This co-experience is usually referred to as *experientiality*—a term made popular by Monika Fludernik. As Toolan (2001: 129) notes, presentation of speech in literary texts contributes to the authenticity of the story world, and Herman (2009: 147) points out that in fiction, "a rich context of felt experience emerges" as a result of "character's conversation." Semino (2004: 436–437) regards character voices as vital to the development of an emotional attachment between the reader and the characters. In fiction, presented discourse can easily project consciousness and help the reader see the events through the eyes of the characters, evoking empathy and sympathy—vital features of emotional engagement according to Toolan (2011). Fludernik (1996) emphasizes that discourse presentation is the basic feature that allows fictional characters to mimic and project human experiences. In other words, introducing speech, thoughts, and writing of characters is what helps make stories fiction. This is particularly true for introducing somebody's thoughts, but I will discuss this issue in the following chapter.

In non-fiction (that is writing that excludes novels, short stories, and other works that are predominately imaginative), including discourse of others serves a purpose different from that of presented discourse in fiction. Semino and Short (2004: 226), having analyzed a large group of fiction and non-fiction 20th-century texts, conclude that fiction relies more on those properties of discourse presentation that emphasize dramatization of the events and inner worlds of the characters. In contrast, non-fiction makes greater use of information-carrying properties of presented discourse such as the ability to summarize.

Semino and Short (2004) do not make it their primary goal to explain why this difference occurs. However, other researchers supply possible explanations. For instance, Livnat (2012) shows that the voices which are introduced into an academic text are entirely subject to the needs of the author and do not contribute to the creation or representation of the identities of those who originated them. Livnat (2012: 64–66) suggests that introducing speech, thoughts, and writing of others serves several specific functions. Firstly, it is the establishing of a research context—that is, showing what other people have said on the subject; secondly, an acknowledgment of the connection with the existing claims/knowledge—showing how the author's work fits with the existing research; and thirdly, the construction of the author's research identity—basically, tell me who you quote, and I will tell you who you are. In other words, presented discourse could be used as a background for the author's ideas.

What is a valued feature of presented discourse in fiction—invitation to the reader to engage with and to interpret the voices of the characters—is to be approached with caution in non-fiction. As Livnat (2012: 59) notes, "In scientific writing, the act of handling other speaker's utterances is less free than in other genres." As a result, when a new voice is introduced, the author, according to Livnat (2012: 59) is obligated to include his/her own interpretation which is to be adhered to by the reader as well. As de Oliveira and Pagano (2006: 641) note, the interpretation of presented discourse supplied by the author contributes to the dialogue between the author and the reader of non-fiction.

The focus on the author as the interpreter of the quoted or paraphrased material can be found outside professional scientific publications as well, in newspaper reports, for example (see Calsamiglia and Ferrero 2003, de Oliveira and Pagano 2006, Smirnova 2009). Presented discourse in such texts is seen as a vehicle for the author's opinions and interpretations not primarily as a mechanism for creating believable characters; outside voices are used to facilitate a dialogue between the author and the reader.

Sometimes, this dialogue is directed at explaining complex information to the reader. Quite often, non-fiction (in an attempt to explain the world) has to introduce multiple voices in such a way that their messages are coherent to a wide audience who might not be well-versed in a particular issue. This is the case with science popularization. In this situation relying on the information-carrying and summarizing properties of discourse presentation accomplishes what Ciapuscio (2003: 210) calls "recontextualizing and reformulating one's source in such a way that it is comprehensible and relevant for a different kind of addressee." It becomes clear that in non-fiction, discourse presentation is used primarily to convey factual information. This function, it appears, dictates the form presented voices are likely to take. For example, as Semino and Short (2004) have found, presented discourse in non-fiction is more likely to be indirect.

To summarize briefly, presented discourse in fiction is directed toward an expression of emotion and dramatization of the described events. Its main goal is to create an emotional response in the reader. Presented discourse in non-fiction is used to boost the accuracy of the account and is directed to the incorporation of facts. In light of this, presented discourse in fiction becomes the means for character creation (see, for example, Toolan 2001), while in non-fiction it is more likely to be the means of constructing the author's professional identity (see, for example Livnat 2012).

When it comes to popular science books, however, the lines are not that clear-cut. Voices of scientists are introduced using the forms most often associated with non-fiction, that is, indirect discourse. At the same time, the functions presented voices perform are the ones commonly observed in fiction, that is, dramatization. This means that the authors of popular science tend to rely more heavily on indirect types of presented discourse but use their summarizing properties to create dialogue and portray scientists as emotionally relatable characters. In other words, presented discourse of scientists is used to fictionalize these texts.

The most striking examples of dramatization, and therefore of fictionalizing popular science, through indirect forms of presented discourse happen when the authors use Narrator's Presentation of Speech Acts (NPSA) to introduce dialogue. While I argue that all types of speech presentation can potentially contribute to dramatization, (Free)Direct Speech and Narrator's Presentation of Speech Acts are the most prominent ways of fictionalizing popular science narratives (see also Pilkington 2018).

(Free)Direct Speech amounts to direct quotations with or without quotation marks or reporting clauses (words of the narrator that introduce the quote). When no quotation marks and no reporting clauses are used, the term Free Direct Speech applies. In the discussion that follows, I will use the term Direct Speech to simplify matters and because all of the examples presented in the chapter are of Direct Speech. Narrator's Presentation of Speech Acts is a paraphrase that indicates that someone spoke but does not give enough details of the message uttered. For more detailed explanations of presented discourse types along with examples see chapter 1.

When discussing Narrator's Presentation of Speech Acts, it is common among linguists to focus on the summarizing properties of this discourse presentation type (see, for example, Toolan 2001, Semino and Short 2004, Short 2007). After all, its main function is to tell the reader that a speech act took place—someone spoke. Yet popular science authors use NPSA to dramatize their narratives. However, dramatization is traditionally associated with direct and free forms of presented speech—quoting—while summarizing functions belong to indirect forms of speech presentation—paraphrasing (see, for example, Semino and Short 2004, Leech and Short 2007).

Such division of functions comes from analyses of primarily fiction texts, and, as a rule, linguists accept the standards developed through examinations of fiction as applicable to non-fiction as well. Short (2007) suggested that Free Indirect Speech (FIS) in fiction may contain dramatizing properties alongside summarizing functions. If an extension of the dramatizing function is possible in fiction, it must be possible in non-fiction as well. Popular science books show ample examples of indirect forms of speech presentation containing dramatizing properties. My findings not only confirm Short's (2007) suggestions but also demonstrate that all types of indirect discourse can carry a degree of dramatization alongside the prototypical summarizing functions.

For example, Narrator's Presentation of Speech Acts are just as common in popular science as Direct Speech—the type of presented discourse usually associated with dramatization. What is even more interesting is that the features of dramatization associated primarily with Direct Speech in fiction—showing of characters' personalities and relationships—in popular science manifest through Narrator's Presentation of Speech Acts. While Direct Speech performs the functions usually associated with it in non-fiction—offering of personal perspective and addition of emotionality.

In a typical novel Direct Speech would be used to introduce character dialogue, and in a typical non-fiction text it would be used to include isolated short quotes. If we are to adhere to these standards, it would be easy to conclude that popular science, being a representative of non-fiction, will not include dialogue. In fact, if you were to flip through a popular science book chosen at random, you will be very unlikely to come across a selection like this:

> "This is not how we want to present ourselves."
> "I know," said Cliff, "I didn't think—"
> "Well, you should have thought about it."
> "It's just a student paper," said Cliff. "And it's the summer weekly issue. Nobody's going to read it."
> "Your interview is in the public record now," Glass snapped.
> "You told me to meet with him!" Cliff burst out.
> "You asked me to speak with him."
> "I assumed that while speaking to Jeff, you'd use your common sense."
> "Look, he asked me about my role in the work. I just answered his questions."
> "Your answers," said Glass, "do not match any of the other stories out there" [Goodman 2006: 167].

This is an example of dialogue using Direct Speech from Allegra Goodman's novel *Intuition*. In a popular science book, if you noticed direct quotations, they probably looked more like this:

> Rous himself later admitted, "I used to quake in the night for fear that I had made an error" [Kean 2012: 140].

But there is dialogue there too. Your eyes are just not trained to spot it because your intuition and experience tell you that dialogue in a printed text looks the way it does in my first example. The dialogue found in popular science, however, is of a different nature—both formally (how it looks) and functionally (what it does). Formally, it is introduced using Narrator's Presentation of Speech Acts and not Direct Speech—that will account for the different look, and functionally, it is often used in order to dramatize not only the immediate events but also to re-imagine scientific debates and present them as dynamic verbal exchanges rather than as a series of publications or talks that took place over a lengthy period of time. In the process of dramatization, the authors project personalities and relationships of the scientists they are writing about. Consider the following example that uses Narrator's Presentation of Speech Acts to create dialogue:

> Miescher protested, but Hoppe-Seyler insisted on repeating the young man's experiments—step by step, bandage by bandage—before allowing him to publish [Kean 2012: 20–21].

This very short exchange gives the reader a glimpse into a conversation between two scientists. There are two speech acts: protested and insisted.

One prompts the other, and they are presented in sequence. The same exchange also introduces hints of the kind of working relationship between these two scientists. We learn that Miescher is clearly a less established figure who needs Hoppe-Seyler's permission to publish his findings. Hoppe-Seyler is the dominant presence in the relationship. At the same time, Miescher's personality is such that he is capable of expressing dissatisfaction to his superior (his speech act is the one of protest), while Hoppe-Seyler's character is revealed as somewhat more careful and not entirely accepting of the younger colleague's success. This is a good example of dramatization, and combined with the summarizing properties of Narrator's Presentation of Speech Acts, it allows the author to keep the story short without sacrificing characterization. Dialogues of this kind are common in narratives dispersed throughout popular science books. Since the narratives themselves are not the end means but rather detours from the main message, keeping their length under control is paramount.

Aside from condensing regular conversations into single sentences, NPSA can be used to explore entire scientific debates and to present them as dynamic dialogic exchanges. In the example that follows, Narrator's Presentation of Speech Acts form the backbone of the interaction; they are responsible to introducing new dialogic turns. Each NPSA is bolded.

Most scientists in the mid–1960s explained the origin of mitochondrial DNA rather dully, *arguing that cells must have loaned a bit of DNA out once and never gotten it back.* But for two decades, beginning with her Ph.D. thesis in 1965, (2) **Margulis pushed the idea that mitochondrial DNA was no mere curiosity….** *Margulis argued, a large microbe ingested a bug one afternoon long, long, ago, and something happened: nothing. Either the little Jonah fought off being digested, or his host staved off an internal coup…. And after untold generations, this initially hostile encounter thawed into a cooperative venture.* (3) **Her opponents countered (correctly) that mitochondria don't work alone; they need chromosomal genes to function,** *so they're hardly independent.* (4) **Margulis parried,** saying that *after three billion years it's not surprising if many of the genes necessary for independent life have faded, until just the Cheshire Cat grin of the old mitochondrial genome remains today.* (5) **Her opponents didn't buy that**—absence of evidence and all—but unlike, say, Miescher, who lacked much backbone for defending himself, (6) **Margulis kept swinging back.** (7) **She lectured and wrote widely on her theory and delighted in rattling audiences.** (She once opened a talk by asking, "Any real biologists here? Like molecular biologists?" She counted the raised hands and laughed, "Good. You're going to hate this.") [Kean 2012: 103–104].

If you read just the bolded parts, you will get a sense of an immediate back and forth. Yet the disagreement and the debate described took years to play out. It is worth noting that NPSA do not work alone to dramatize this debate. They are supplemented by Indirect Speech (italicized), Free Indirect Speech (italicized and underlined), Direct Speech and Narration, more specifically, commentary from the author (underlined). The NPSA supply the outline with Direct Speech, Indirect Speech, Free Indirect Speech, and author commentary providing the details and expanding the ideas introduced by NPSA. Each dialogic turn, in fact, begins with a Narrator's Presentation of Speech Acts.

While all of the forms of speech presentation found in this example work together to dramatize the event, each type retains the properties common for it in popular science. Since Narrator's Presentation of Speech Acts are most often associated with presentation of dialogue, they guide the exchange. Indirect Speech, introduces scientific hypotheses. The role of Free Indirect Speech in this exchange is to introduce a metaphorical explanation of the scientific hypothesis—a role often observed in popular science texts. Direct Speech, as is common, appears in a form of a short quote designed to inject emotionality and personal perspective.

Now, let's look at each type of speech presentation in detail. The hypothesis for the narrative is presented in the form of an argument to keep in line with the general concept of the scientific debate presented as a dialogue. The first occurrence of Indirect Speech is introduced by the reporting verb "argued." The second occurrence uses a neutral reporting verb "saying"; however, it follows an NPSA that positions the hypothesis as a counter argument, "Margulis parried."

The verbs used in NPSA project confrontation: "pushed," "explained," "countered," "parried," "didn't buy," "swinging." All of them except "explained" are used in their metaphorical senses since their literal meanings indicate physical rather than verbal actions, and verbal actions associated with physical confrontation at that. As parts of NPSA the verbs listed above embody the dramatizing properties, as they present the argument in terms of a physical fight. The fact that Narrator's Presentation of Speech Acts introduce aggression, allows for reporting verbs used in Direct Speech and Indirect Speech ("saying," "asking") to remain neutral.

Free Indirect Speech supplies creative explanations, which are attributed to the scientists. Considering the first instance of FIS ("Either the little Jonah fought off being digested, or his host staved off in internal coup"), it is unclear if Margulis herself employed the Biblical reference, or whether it is coming from the author, but the use of Free Indirect Speech attributes the metaphor to the scientist. The same happens with the second occurrence of FIS ("until just the Cheshire Cat grin of the old mitochondrial genome remains today"). It also uses a metaphor that follows the hypothesis. It is possible to see these portions as narration since they appear out of place in a scientific debate. However, they are different from the instances of narration (underlined). Narration in this case tends to interrupt not explain.

Direct Speech, like Indirect Speech and Free Indirect Speech, is also used to expand the NPSA it follows. The direct quote shows an interaction outside the immediate dialogue ("a talk"), but the lecture Direct Speech exemplifies serves as an example and a continuation of the larger debate projected via Narrator's Presentation of Speech Acts.

Overall, this example demonstrates that dramatization can be achieved not through one or another single type of speech presentation but through a combination of several types, each of which possesses dramatizing properties of some degree. This includes indirect forms of presented discourse such as Indirect Speech, Free Indirect Speech, and, of course, Narrator's Presentations of Speech Acts. In the process, those types of speech presentation that are not commonly associated with dialogue in popular science (DS, IS, FIS) do contribute to dialogic exchanges. In fact, they play vital roles by introducing the subjects of the debates—the hypotheses—and supplying examples and explanations. For instance, Indirect Speech gives more details to the preceding NPSA. Free Indirect Speech that follows Indirect Speech supplies explanations for the latter. The level of emotionality (aggression, in this case) decreases with each subsequent discourse presentation type. Narrator's Presentation of Speech Acts introduce the argument in the most aggressive manner using the verbs that indicate a physical confrontation, while Indirect Speech and Direct Speech use neutral reporting verbs, and Free Indirect Speech does not present arguments at all but provides explanations. Working together, the multiple types of speech presentation produce a detailed dramatization of the events. At the same time, by making NPSA the backbone of the dialogic exchange, the author condenses a debate that took decades into a series of dynamic exchanges each of which is a reaction to a previous statement. This is an example of the summarizing and the dramatizing functions of NPSA working together and being supplemented by other forms of presented speech.

In addition to dramatizing events, Narrator's Presentation of Speech Acts are also used to introduce and/or explain scientific phenomena. It is possible to divide the non-dramatizing NPSA into two groups: those that present scientific concepts and ideas and those that showcase the issues outside science that nevertheless influence it and present a personal perspective of the scientists on the subject discussed. Most of the NPSA belong to the first group. Here is an example:

> On his sixty-fourth PowerPoint slide, Incandela revealed what you get when you combine these two channels together: 5.0 sigma [Carroll 2012: 184].

Such Narrator's Presentations of Speech Acts hardly explain anything; they simply present scientific content. These NPSA supply a summary which is elaborated on using either Narration or other types of presented speech, most often Free Indirect Speech. Consider an example (NPSA bolded, narration underlined):

> **Italian physicist Dario Autiero announced a result that ended up being more infamous than famous: neutrinos that appeared to be moving faster than the speed of light.** <u>The finding came from the OPERA experiment, which tracked neutrinos that were produced at CERN and traveled 450 miles underground to a detector in Italy. Because neutrinos interact so weakly, they can pass through many miles of solid rock with very little loss of intensity, making this kind of arrangement a uniquely effective window onto their properties</u> [Carroll 2012: 195–196].

In the next example, NPSA are working together with Free Indirect Speech (NPSA bolded, FIS italicized):

> **Some argued that Rous had misdiagnosed the tumors**; *perhaps the injections caused an inflammation peculiar to chickens* [Kean 2012: 140].

When used in narratives of discovery, NPSA followed by narration usually appear at the beginnings of stories—introducing the discovery. Those NPSA that are followed by Free Indirect Speech are more likely to be found in the middle of a discovery narrative.

The second, less numerous group of Narrator's Presentation of Speech Acts that do not create dialogues tie scientific issues they introduce to social or historical events. Consider this example:

> The American scientific community, which was still grappling with fusion in the fierce race with Russia to produce the H-bomb, declared that the claim was nonsense [Kaku 2011: 236].

Note the reference to the arms race with Russia and the emotional renunciation of the claim (underlined). There is more going on than a simple introduction of scientific facts. Science is shown in a social and historical context.

While NPSA can perform various functions in a popular science text, their main role is to dramatize the events and to create dialogue-like exchanges. In non-fiction, these tasks are usually associated with Direct Speech. However, in popular science, direct quotes from scientists are used for other purposes.

Those who study the role of direct discourse presentation in non-fiction single out the following functions most commonly performed by Direct Speech: distancing the author from the presented claim (drawing a clear line between his or her own words and those of the quoted experts;

this allows for deniability should the claim be proven faulty or problematic in the future), establishing credibility and reliability (quoting experts to show support for the author's own ideas), supplying accuracy (direct quotes are always more reliable sources of information than paraphrases), and providing personal perspective of the original speaker (using the speaker's own words allows for the introduction of unique vocabulary or point of view). Dramatization is not on the list. That is because, it is usually assumed, that Direct Speech reveals its dramatizing properties in fiction only. However, some studies demonstrate that DS retains its dramatizing functions in non-fiction. In Bell's (1991: 209) words, Direct Speech adds "a flavour" of the original speaker's "own words." It also, according to Semino and Short (2004: 95), may be used for "dramatizing protagonists' lives." When analyzing the functions of DS in non-fiction, Bell (1991), Calsamiglia and Ferrero (2003), and Semino and Short (2004) draw on examples from newspapers and (auto)biographies. Their findings on the dramatizing properties of DS can be corroborated through analysis of popular science as well. At the same time, it is fair to say that in popular science, the dramatizing properties of DS are limited. Direct Speech in popular science covers only some aspects of dramatization, namely emotionality and personal perspective. There are clear examples of both. Let's look at emotionality first:

> "What a field of novelty is here opened to our conceptions!" he [Herschel] exclaimed, more delighted by the variety of the sky than bothered at having been wrong [Ferris 1988: 157].

This particular direct quote clearly shows Herschel's emotional state as he discovered that nebulae can be made of gas as well as of stars. Now let's consider "a flavour," to use Bell's (1991) word, of the scientists' speech, which is introduced through Direct Speech representing personal perspective:

> "Well boys, we've just been scooped," he [Dicke] told his colleagues as he hung up the phone [Bryson 2003: 12].

The direct quotation in this example is used to expose the reader to Dicke's ways of speaking around his colleagues—the reader is to believe that Dicke referred to his fellow researchers as "boys" and used the verb "scooped" to indicate that Penzias and Wilson beat them to the discovery of the cosmic microwave background radiation.

The two examples above also illustrate important differences in how direct voices of scientists are employed by popular science authors. For Direct Speech that introduces emotions, the emotionality can be projected in two ways: either in the reported clause (the actual quote) or in the reporting clause (the author's introduction). It is also possible to have both the reported and the reporting clauses contain emotionality.

In the example about Herschel, the emotionality is projected as coming from both the scientist (use of an explicative and an exclamation mark) and the author (emotionally evaluative reporting clause—"he exclaimed"). When the author injects his own emotional evaluation of the quote in the reporting clause in combination with emotionally charged words of the scientist, that author appears to reinforce the emotionality of the utterance.

In some instance, on the other hand, the only emotionality associated with DS comes from the reporting clause. That is, the author is projecting his own emotions, while the quote is emotionally neutral:

> Weber was so excited by the potential of their discovery that he prophetically declared, "When the globe is covered with a net of railroads and telegraph wires, this net will render services comparable to those of the nervous system in the human body, partly as a means of transport, partly as a means for the propagation of ideas and sensations with the speed of light" [du Sautoy 2011: 177].

When Weber's utterance is evaluated without du Sautoy's introduction, there is no expression of emotion. In fact, the reader would not know of Weber's excitement, nor would he/she necessarily feel excited about the information. However, the author's emotionally charged introduction claims that the scientist produced these words as an exclamation of excitement upon the discovery of how to transmit messages via electric wires. The adverb "prophetically" reinforces the importance assigned to the discovery. In choosing to use an emotionally charged reporting clause, du Sautoy is guiding the reader's emotional reaction to Weber's words, making sure that the reader not only understands the practical significance of science but also experiences the rise of emotions associated with discoveries.

It is noteworthy that the emotions reporting clauses project, and therefore the emotions that the reader is to experience, are always positive. They are emotions of excitement, enthusiasm, and, at times, surprise. Negative evaluation of discoveries, if introduced, always comes from the fellow scientists, not from the authors. Calsamiglia and Ferrero's (2003) findings suggest that by using reporting clauses the authors can either predispose or turn away the readers from the presented voices in the text. Clearly, the authors of narratives of discovery want their readers to appreciate the scientists they write about and their achievements. By increasing the emotionality of discovery accounts, especially when that emotionality is projected by the authors, Direct Speech inevitably fictionalizes the texts. Showing scientists as expressing emotion goes beyond the stereotypical portrayal. The authors are clearly trying to project characters that are relatable yet realistic.

Even when DS is not projecting emotionality, it is still used to helps transform scientists from impersonal researchers into human characters. The authors achieve this effect by employing Direct Speech that introduces personal perspective. Let's look at a couple of examples:

> Lahn screened different populations alive today and determined that the brain-boosting versions [of microcephalin and aspm genes] appeared several times more often among Asians and Caucasians than among native Africans ... follow up studies determined that people with these genes scored no better on IQ tests than those without them. Lahn ... soon admitted, "On the scientific level, I am a little bit disappointed. But in the context of the social and political controversy, I am a little bit relieved" [Kean 2012: 344–345].

Lahn's Direct Speech is not necessarily emotional; its purpose is to show the scientist's personal reaction to the findings that have proven him wrong.

Personal perspective of scientists allows the readers to experience events not through the interpretation offered by the author or through their own assumptions but through the first-hand accounts. This use of Direct Speech might not always show the members of the scientific community as flawless, but it does make them into interesting characters with genuine human qualities.

In addition to using Direct Speech and Narrator's Presentation of Speech Acts, popular science authors also rely on Indirect Speech to incorporate the voices of the scientific community into their texts. As I mentioned earlier, Indirect Speech is common in those parts of texts that include explanations of scientific concepts. The most striking feature of Indirect Speech in popular science is its ability to explain science through the use of figurative language: metaphors, analogies, and prosopopoeia. The last instance of figurative language refers to giving inanimate objects (like articles or papers) an ability to speak. Consider an example of each:

> Metaphor: The authors concluded that we are all immersed in a bath of photons, a cosmic heirloom bequeathed to us by the universe's fiery birth [Greene 2011: 39].
>
> Analogy: It was, he [Rutherford] remarked, as startling as if a bullet were to bounce off a sheet of tissue paper [Ferris 1988: 256].
>
> Prosopopoeia: A prescient series of papers by the German mathematician Theodor Kaluza and by the Swedish physicist Oskar Klein suggested that there might be dimensions that are proficient at evading detection [Greene 2011: 84].
>
> Kaluza-Klein [meaning the theory] echoed across the decades answering that the dimensions are all around us but are just too small to be seen [Greene 2011: 88].

Traditionally, Indirect Speech (which is essentially a paraphrase) has been valued for its ability to present the message but omit its exact wording—

a potentially useful aspect when it comes to popularization since not every word uttered by scientists can be clear and straightforward enough to be incorporated via direct quotations. Some of what scientists say even in discussions with non-specialists, may be too technically worded to be quoted directly and requires reformulation (see Ciapuscio 2003). Indirect Speech supplies an excellent means of reformulation, and figurative language adds creativity and imagination to passages that otherwise might appear too technical and boring.

The use of figurative language in Indirect Speech raises an interesting question: whose creativity and imagination is being showcased? Is it the author who is being creative? Or the scientist to whom Indirect Speech is attributed? This is especially relevant for metaphors and analogies since they are parts of IS itself rather than of the reporting clause as is the case with prosopopoeia. When we remember the primary function of Indirect Speech—content presentation—the conclusion is that IS reflects only the ideas and not the original words. At the same time, the creative presentation of a scientific issue is a conscious decision and thus could be regarded as part of the content. This line of thought would suggest that analogies and metaphors found in Indirect Speech should be attributed to the scientists. On the other hand, the reformulating properties of IS dictate that the authors are the more likely originators of figurative language in order to explain the material and make it more relatable. I tend to think that the second proposition is closer to the truth. Let me explain why.

First of all, popular science authors capitalize on the freedom to reshape the wording of utterances that Indirect Speech provides and choose to restate the original in a simpler and a more engaging way. The example above that illustrates a metaphor (Greene 2011: 39) is a summary of the result found in "the papers of Gamow, Alpher, and Herman that in the late 1940s announced and explained" the cosmic microwave background radiation—a text which might pose difficulties for a non-specialist. By using IS instead of quoting from the conclusion verbatim, Greene seizes the opportunity to introduce a metaphor. He is not simply offering a more tolerable explanation; along the way he is creating a sense of the sublime: "a bath of photons, a cosmic heirloom bequeathed to us by the universe's fiery birth."

Turney (2004a: 91) defines the sublime as "an aesthetic category," with the help of which "Science writers evoke their most telling effects." He goes on to say that "the feeling generated by the sublime includes both awe at the overwhelming sensory impact … and at the human capacity to apprehend it in its full extent" (Turney 2004a: 93). Turney (2004a: 93) also suggests that there could be a more practical application for the metaphors of the sublime since they can function as "the safeguard against the feelings of insignificance induced by cosmic immensities." The fact that the authors choose to incorporate the sublime as part of presented discourse points to a conscious decision to associate the advantages of this strategy (the ability of the human mind to comprehend the universe, anticipation of the possible feelings of insignificance on the part of the reader) with the scientists and thus portray them in a positive light. In other words, the authors give credit to the scientists for especially evocative passages even when it is not due. This is a smart strategy as their goal is to create characters who come through as affable, intelligent, and caring.

The use of certain imagery associated with the sublime presents another point in favor of the authors' creative powers and not necessarily the poetic diction of the scientists. As Turney (2004a: 90) notes, "The sublime has become the characteristic aesthetic of much contemporary popular science." Metaphors of the sublime have been employed by such celebrity popularizers as Carl Sagan and by the relatively new authors such as Adrian Woolfson. Turney (2004a) argues that the phenomenon has been used so often that certain standards have emerged. For example, he identified "the vocabulary of the sublime" as predominantly seascapes and landscapes (Turney 2004a: 96). The use of figurative language is so ubiquitous in popular science that certain metaphors and analogies start to contribute to a pool of stock imagery that many authors "adopt and modify" for their own purposes (Turney 2007: 2). In contrast, new metaphors, according to Turney (2004b: 337), indicate that "there is not yet a widely accepted formula for describing ... novel" ideas. Turney (2004b: 343) suggests that the adaptation and modification of certain metaphors indicate the success of their originator. If someone uses a particularly apt metaphor or analogy once or twice, it becomes fair game for other authors in the popular science community. The example of an analogy that I introduced earlier in this section and will repeat below contains an image that is also found in Bryson (2003). Here is the example quoted earlier:

> It was, he [Rutherford] remarked, as startling as if a bullet were to bounce off a sheet of tissue paper [Ferris 1988: 256].

Here is Bryson's (2003: 139–140) take on the same incident:

> It was as if, he [Rutherford] said, he had fired a fifteen-inch shell at a sheet of paper and it rebounded into his lap [Bryson 2003: 139–140].

Neither Ferris (1988) nor Bryson (2003) cite the source for Rutherford's Indirect Speech. And in light of Toolan's (2001: 128) note that "People are quite capable of 'reporting' things that their reportees never said," it is possible to assume that Rutherford was never inspired to verbal creativity by the failure of his experiment. Bryson (2003: 5, 6) in the introduction to his book makes it pretty clear where he got the metaphor. Bryson (2003: 6) acknowledges that he collected his material by "reading books" among other means, and he mentions Ferris by name as an example of a science author who writes "the most lucid and thrilling prose" (Bryson 2003: 5).

Indirect Speech, as it is used in popular science, presents some of the most striking and beautiful explanations of scientific phenomena. In some cases of figurative language that appears in IS we might never be sure who to give credit to, but this is not the point. The main idea is the willing and deliberate attribution of creativity to the scientists. Each of the authors who uses IS with metaphors and analogies could have easily expressed the same content using narration and taken all the glory (neither one of the examined books cites sources in cases of IS), but by opting out for IS these authors are letting the scientists shine.

The most basic function of speech presentation in popular science amounts to dramatization. This is somewhat unusual for this genre. Traditionally, dramatization is not the most prominent function of presented discourse in non-fiction. Popular science, however, in an attempt to introduce scientists as relatable and likable characters uses the resources available to non-fiction to produce effects most common in fiction texts. That is how we end up with indirect discourse that creates dialogue (Narrator's Presentation of Speech Acts) or produces figurative explanations (Indirect Speech). Direct Speech—the type of speech presentation most often associated with dramatization—in popular science is employed mainly to introduce short quotations that add personal perspective or emotional assessment. These functions are not unusual for DS in non-fiction.

Overall, the authors of popular science take the conventional resources of non-fiction (indirect discourse) and use them to infuse their stories of science with fictional elements (Pilkington 2018).

6

What They Imagine Is Possible

Thoughts of Scientists

Those who study presentation of speech, writing, and thought often point out the different roles each type of discourse presentation plays. The differences are said to be especially striking when speech and thought presentation are compared. For instance, Semino and Short (2004: 118) confirm that "the effects that result from their [thought presentation] types are quite different from those we have noted for speech and writing." The difference, as Leech and Short (2007: 270) point out, lies in the inaccessibility of thought: "We cannot see inside the minds of other people." Short's later works (see, for example Short 2007 and Short 2012) further this point by stressing that thought presentation does not possess those communicative properties that are common for speech and writing presentation.

Such view of thought presentation leads to a popular assumption that since thought is not directly accessible in everyday life and is not used for communicative purposes, presentation of thought centers on the inner worlds of characters (see, for example, Toolan 2001, Semino and Short 2004, Leech and Short 2007, Short 2007, Short 2012). As a result, thought presentation is usually discussed in close connection with dramatization. This is a perfect perspective for my argument that presented discourse in popular science serves to dramatize the accounts of scientific events. However, popular science, it turns out, is very different from other genres, and the standards set for them (especially standards based on analyses of fiction) do not always work.

Thus the functions of presented thought in popular science are radically different from those previously observed in other genres. Instead of showing the reader the inner-most secrets of the scientists, the authors of popular science use thought presentation to introduce scientific hypotheses and discoveries. By doing so the authors of popular science take thought presentation out of the realm of the intimate. They use thoughts of scientists to introduce the reader to common knowledge or cutting edge scientific advancements rather than to intimate thoughts. In other words, the messages delivered through presentation of thought are very much public and, for the most part, do not contribute to characterization of the scientists, something that would be expected of thought presentation based on the previous analyses.

Thoughts of scientists as presented in popular science are closely associated with the experimental and the empirical (and thus the more physical rather than the mental) sides of science. The decision to focus on the empirical side of science through the presentation of mental processes might be an attempt at establishing a kind of closeness between the reader and the scientific issues discussed. On the other hand, the introduction of scientific information through such an intimate means of discourse presentation points to the idea that scientific progress is a result of individual effort rather than the collaboration among the members of the scientific community. Thought presentation in popular science might not contribute to character creation as directly as it does in fiction, but it does place the focus on individual scientists just the same. As Parkinson and Adendorff (2004: 388) point out, popular science is concerned with "people and what they say and think." Therefore, it is important to present scientific results (discoveries) as originating with specific scientists. This focus, in many ways, undermines the complexity of the discovery process (for more on this see Pilkington 2017).

As a general rule, experimental procedures are not foregrounded in popular science, and the discoveries are often described as products of intellectual rather than experimental processes, so it appears that overall scientific discovery is a result of thinking rather than doing. By introducing thoughts of scientists, the authors bring the experimental procedures to the attention of the reader. The authors do not always take it upon themselves to give accounts of experiments. That is why narrated segments often skip the technical details (for more on this see chapter 4). Until we look at the presented thoughts of scientists, we do not get a clear sense that experimentation is an important aspect of scientific endeavors. Overall, presented thoughts are dedicated not so much to the unveiling of the inner worlds of scientists but rather to tracing the mental processes and reactions to empirical work which result in discoveries. Narrator's Presentation of

Thought Acts (NPTA) and Indirect Thought (IT) aid the most in accomplishing this goal. These two types of presented discourse are responsible for introductions of scientific hypotheses and discoveries in popular science.

Indirect Thought and Narrator's Presentation of Thought Acts are the least dramatic means of thought presentation. For instance, Toolan (2001: 139) writes, "recourse to more direct thought-presentation than IT [Free Indirect Thought, Direct Thought, Free Direct Thought] may ... invite the inference of ... 'entering' of the character's intimate mental space." Like Toolan (2001), Semino and Short (2004: 128, 131) observe that IT and NPTA are "less dramatic" means of accessing the inner world of a character. Semino and Short (2004: 115) note the relative lack of Narrator's Presentation of Thought Acts across the texts they analyzed, especially in non-fiction. The lack of NPTA in non-fiction texts other than popular science suggests that even in non-fiction authors turn to presentation of thought as a means to dramatize; that is, to introduce the intimate sides of the people they write about.

In popular science, the preference for the most non-dramatizing categories of thought presentation falls in line with the purposes of IT and NPTA—83 percent of all thought presentation occurrences in the narratives of discovery are devoted to the introductions of scientific hypotheses and breakthroughs.

The introductions of scientific hypotheses and discoveries reveal several important points about thought presentation in popular science. Firstly, they demonstrate a strong connection between presented thoughts of scientists and narration that describes experimental procedures, thus supplying evidence for thought presentation being concerned more with the empirical than with the intimate. Secondly, presentations of discoveries and hypotheses follow specific verb patterns that generalize thought presentation and, again, provide support for my argument that thought presentation does not focus on expressions of individual inner worlds. In fact, the verb choices I identified during the analysis of hypotheses and discoveries function as what Mildorf (2008: 288) calls "a mitigating strategy which helps the speaker disclaim any ultimate knowledge or access to ... other people's minds." That means that certain verbs used to introduce thoughts of scientists (verbs in reporting clauses) indicate that the information that follows is not necessarily an expression of someone's thoughts but merely an idea that the author chose to express as such. Other than emphasizing scientific issues, thought presentation is also responsible for contributing to a positive image of scientists through highlighting hypotheses that have been proven correct. Let us consider a few examples.

6. What They Imagine Is Possible   109

Scientific hypotheses are an important part of popular science books; they carry special significance in narratives of discovery. Quite often, they are signposts that guide the readers' expectations of what is to be discussed. About 20 percent of the narratives of discovery use presented discourse to introduce hypotheses, and these narratives rely overwhelmingly on presentation of thought to do so. Over 40 percent of thought presentation in the narratives of discovery is used to present scientific hypotheses while only 13 percent of speech presentation is employed in this way. The example below demonstrates a hypothesis presented via Indirect Thought:

> Many scientists at the time were skeptical, but Shope wondered if rabbit "horns" were also tumors, somehow triggered by an unknown virus [Zimmer 2011: 24].

This use of Indirect Thought is typical for introducing ideas that later on lead to discoveries. Like all hypotheses expressed through thought presentation, the current example contains a verb of mental action ("wondered") and a hedge ("if") to indicate uncertainty that is to be eliminated once the proof is obtained.

In addition to using Indirect Thought, the authors also rely on Narrator's Presentation of Thought Acts to introduce hypotheses. Consider the following example:

> This set Bunsen to wondering whether they might be able to detect chemical elements in the spectrum of the sun as well [Ferris 1988: 164].

At the same time, it is also common for NPTA to work alongside narration and presentation of speech to produce texts segments that reveal scientific hypotheses. Narrator's Presentation of Thought Acts that appear in combinations with speech presentation and narration do not themselves introduce hypotheses and are more likely to resemble prototypical NPTA observed by Semino and Short (2004: 130) and described as "occurrences of a specific individual thought in the mind of a participant in the story, which do not include any indication of the propositional content or the 'wording' of the thought." Such Narrator's Presentation of Thought Acts, as Semino and Short (2004: 131) suggest, most often introduce the character's motivations that help explain his/her speech or actions that precede or follow. In popular science, the thoughts and motivations that Narrator's Presentation of Speech Acts help explain are inevitably connected with the scientific ideas. Consider an example (NPTA is bolded followed by the italicized presentation of speech):

> Thompson and his colleagues at the Cavendish Laboratories began to measure the electrical charge and the weight of some of these radiations. **They tried to decide how these two measurements were related to each other.** In 1987 *Thompson proposed that these rays were streams of charged subatomic particles: bits of atoms* [Bynum 2012: 183].

While the NPTA does not introduce the hypothesis directly, it explains the reasoning that went into the well-formulated utterance given via Indirect Speech (in italics). What is noteworthy about interactions of discourse presentation types in hypotheses is that the different types of discourse presentation introduce the different chronological stages of the discovery process.

While combinations of discourse presentation types and narration do play their role in the introduction of hypotheses, it is more common to use one type of thought presentation to introduce scientific ideas. In fact, single-occurrence hypotheses presentations reveal clear verb choices associated with hypotheses expressed through Narrator's Presentation of Thought Acts and Indirect Thought. Look at the following example:

> He wondered if something other than bacteria was responsible for tobacco mosaic disease, something far smaller [Zimmer 2011: 4].

The hypothesis is introduced via Indirect Thought. The verb "wondered" is part of a pattern for hypotheses presentation via scientists' thoughts. There are four verbs/verbal phrases that occur most often with hypotheses in presented thoughts: "wonder," "come up with an idea," "think," and "assume." The first two most often indicate what I call "positive hypotheses"—hypotheses that are proven correct as the story progresses. The last two are associated with "negative hypotheses"—hypotheses that are later refuted. Accordingly, the example given above introduces a positive hypothesis, while the example that you are about to read presents a negative hypothesis:

> Perhaps, he thought, the plants were suffering from an invisible infection [Zimmer 2012: 3].

Other than the verb used to present it, it is impossible to distinguish a negative hypothesis from a positive one without the larger context. The verb choice, however, creates a distinction between the two hypotheses and can help predict the narrative's resolution when hypotheses are analyzed in isolation. The two examples are taken from the same narrative about the discovery of viruses and show that other than the verb used in the reporting clause of Indirect Thought, there is nothing to separate a negative hypothesis from a positive one. For instance, both contain hedges ("if" and "perhaps") that point to the tentativeness of the statements. This demonstrates the key role reporting verbs play in thought presentation introducing hypotheses.

It is significant that positive hypotheses outnumber the negative. The emphasis on the positive hypotheses confirms the general focus of popular science books on the successful outcomes of science. If we look at the narratives of discovery told in popular science books, majority of them will be stories with favorable outcomes which somehow contribute to either scientific progress or technological advancement of the human society. Very few narratives describe what in chapter 4 I called "failed discoveries." Negative hypotheses, however, are much more likely to appear not as focal points in the narratives of failed discoveries but as side steps in the narratives of success.

After hypotheses presentation, the introduction of discoveries forms the second most numerous functional category of thought presentation. Indirect Thought appears to be the only type of thought presentation for signaling these important points. (Other options include narration and some instances of Narrator's Presentation of Speech Acts.) Consider an example:

> An atom, Rutherford realized, was mostly empty space, with a very dense nucleus at the center [Bryson 2003: 140].

This example demonstrates the most popular way of incorporating statements of discoveries into larger text segments. Such instances of Indirect Though use pragmatic as opposed to figurative language. The descriptions of thoughts utilize generic terms of the disciplines to present the discoveries. This method reveals that the authors prefer certain verbs when using Indirect Thought to introduce discoveries. In other words, Indirect Thought that presents discoveries acts similarly to hypotheses-presenting Narrator's Presentation of Thought Acts.

Almost all of the discoveries introduced via Indirect Thought use the verb "realize." The other preferred verbs are "occur" and "assume." Except for some instances that use the verb "assume" (e.g., "she assumed ... that the change involved hydrogens shifting around" [Kean 2012: 100]), majority of the discoveries introduced through IT contain an element of sudden enlightenment, which the verbs "realize" and "occur" express. At a first glance this observation suggests that some scientific discoveries are being presented as serendipitous insight not necessarily dependent on consistent empirical work. However, when the instances of discoveries introduced via IT are examined in larger contexts, the connection between thought presentation and descriptions of experiments provided via narration show that just the opposite is true.

By far, most of the discoveries expressed by Indirect Thought are preceded and/or followed by descriptions of experimental procedures, presenting the Eureka moment as a reaction to a specific experimental result.

Here is the continuation of the story from the previous example about Rutherford:

> In 1910, Rutherford ... fired ionized helium atoms, or alpha particles, at a sheet of gold foil. To Rutherford's astonishment some of the particles bounced back. *It was as if he said, he had fired a fifteen-inch shell at a sheet of paper and it rebounded into his lap. This was just not supposed to happen.* **After considerable reflection he realized there could be only one possible explanation: the particles that bounced back were striking something small and dense at the heart of the atom, while the other particles sailed through unimpeded. An atom, Rutherford realized, was mostly empty space, with a very dense nucleus at the center.** This was a most gratifying discovery, but it presented one immediate problem. By all the laws of conventional physics, atoms shouldn't therefore exist [Bryson 2003: 139–140].

In this context, the discovery presented as Indirect Thought (bolded) does not seem as sudden or as unfounded as it might when IT is analyzed in isolation as in the example above. This particular example is also a good showcase for the interactions of presented discourse and narration. (Indirect Speech and Free Indirect Speech are italicized; narration is underlined.)

The narrated segment supplies the details of the experiment (and later the evaluation of the discovery); Indirect Speech and Free Indirect Speech show Rutherford's reaction to the experiment, and the second sentence of Indirect Thought presents the discovery itself as a culmination of all the previous activities described using discourse presentation. The combination of presented discourse and narration works to create a chronological account of the discovery and to position it as the outcome of the experiment. Rutherford's thoughts are presented as focused on the observation, from which he deduces the structure of the atom—his discovery.

When Indirect Thought that is used to introduce a discovery is not connected with an experimental procedure, the discovery process appears underdeveloped, and the discovery announcement comes somewhat suddenly. However, this mode of introducing discoveries seems to be a deliberate choice on the part of the authors. A discovery presented through Indirect Thought which is not connected to an experiment is usually not the main discovery of a narrative, or it mimics the actual events where a discovery was indeed a sudden realization. The two examples that follow illustrate both scenarios. In each case, Indirect Thought is bolded:

> Despite the evidence mounting against him, Bekenstein had one tantalizing result on his side. **In 1971, Stephen Hawking realized that black holes obey a curious law** [Greene 2011: 247].

Hawking's discovery helps Bekenstein strengthen his own theory and provides evidence for the main discovery of the narrative—the multiverse. There is no need to take up room with the descriptions of Hawking's discovery process; the mention of the discovery is sufficient. The remainder of the narrative provides a brief explanation of Hawking's findings and connects them to Bekenstein's ideas.

> As often happens, the answer came to him not while he was at work in his observatory but while he was relaxing. While on a boat in the Thames, Bradley found himself gazing at a wind vane mounted atop the mast. It pointed into the wind and therefore seemed to change direction whenever the boat turned. What was changing, of course, was the orientation, not of the wind, but of the boat.
> **It occurred to Bradley that the earth is like a boat adrift in winds of starlight—that, as the earth moves through the starlight, its motion alters the apparent positions of the stars** [Ferris 1988: 138].

This example presents a different reason for not including a description of the experiment—it did not happen—and illustrates the use of figurative language in the description of the discovery. Ferris (1988: 137–138) tells a different discovery story, but one that nonetheless is not uncommon, as he himself suggests.

Thought presentation is usually regarded as arbitrary since it is impossible to trace it back to an actual utterance—we keep our thoughts to ourselves, and once we voice them, they become speech and no longer thoughts. That is, no matter what we say we think, no one can check if it is true (for more on this see Short 2012: 23). That means that the criterion of faithfulness is even less applicable to thought presentation than it is to presentation of speech or writing. This is what leads some researchers (see, for example, Cohn 1990: 784–785, Dawson 2015: 80) to suggest that if at any time presentation of thought occurs without specific references to memoirs, journals, or similar materials, it is fictional and therefore fictionalizes a text in appears in. This line of argument gains more strength once thought presentation becomes detailed and includes figurative language as is the case with the last example. The analogy between the boat on the water and the earth in space is clearly attributed to Bradley, but it is not clear if the scientist expressed it in quite the same creative manner as Ferris (1988: 138) did.

While I do not subscribe to the line of argument that suggests any manifestation of thought presentation automatically designates a text as a fiction, I still believe that more elaborate instances of presented thought contribute to dramatization and thus do fictionalize the texts in which they appear. In general terms, thought presentation is usually discussed in close connection with dramatization. For instance, Toolan (2001: 139) and Semino and Short (2004: 123, 128, 131) explain the differences in the effects produced by thought presentation in relation to the degree of dramatization.

Semino and Short (2004: 121, 123) also note that there is not much difference in the effects produced by thought presentation in fiction and nonfiction: in both thought presentation is connected with dramatization, differing only in the degree. Thus Direct Thought, Free Direct Thought and Free Indirect Thought (the types of thought presentation considered as possessing the highest dramatizing properties by Toolan and Semino and Short) are more common in fiction, while almost nonexistent in nonfiction.

The scarcity of these forms in popular science books supports the claim I made at the beginning of this chapter that the functions of thought presentation lie beyond dramatization. The example from Ferris (1988: 138) is an anomaly because it uses figurative language. The overwhelming majority of occurrences express the thoughts of scientists in more pragmatic language less likely to raise questions of faithfulness. In general, the lack of elaborate descriptions in thought presentation and the preference for reporting verb patterns (certain verbs associated with positive and negative hypotheses and almost universal use of the verb "realize" for the presentations of discoveries) constitute the mitigating strategies that suggest presentation of thought in popular science is used for communicative purposes rather than for revealing the intimate inner worlds of scientists. At the same time, the few instances of figurative language in thought presentation show the willingness of the authors to present scientists as capable of creative approaches to their work. Figurative language is, however, not common in presentation of thought, but it is more likely to occur in Indirect Speech (see chapter 5).

Overall, presentation of thought in popular science functions in unexpected ways. It is used primarily to introduce scientific hypotheses or discoveries rather than give the reader insight of scientists' private worlds—a kind of intimate access common in fiction and even in non-fiction, as previous studies show. In popular science, as we have seen, thought presentation is used differently—to announce discoveries and introduce scientific hypotheses. In general, thought presentation is more science-oriented than presentation of speech, which tends to highlight personal relationships and arguments (Pilkington 2018).

Presentation of writing shows a whole different side of the voices of scientists—it is the subject of the following chapter.

7

Literature and Limericks
Writing in Popular Science

So far, I have introduced presentation of speech and presentation of thought as the ways in which popular science authors incorporate voices of scientists into their texts. These two categories are the classics of presented discourse analysis, and until 2004 they remained the only ones available. Semino and Short (2004), after a detailed investigation of 120 text samples ranging from novels to newspapers to autobiographies, noticed that relaying what someone said or thought were not the only options to introduce an outside voice into a text. People also write. And sometimes, authors choose to introduce instances of writing as a means of characterization or a way to deliver information more accurately.

Here is what Semino and Short (2004: 111–113) say about presentation of writing: "The use of direct quotation from written sources normally imposes higher faithfulness constraints than from spoken sources." The whole idea behind the creation of writing presentation was accuracy—Semino and Short (2004: 48) note, for example, that Direct Writing may create a more "accurate word-by-word representation" than Direct Speech.

This would imply that texts more concerned with accuracy would choose presentation of writing in favor of presentation of speech or thought. Indeed, Semino and Short (2004: 113) discovered that "categories of writing presentation are slightly more frequent in the serious sections" of the texts they examined. To extrapolate this conclusion to popular science books (which I would consider serious texts) would be to expect a lot of instances of writing presentation. It is not the case, however. Let me offer a few suggestions why.

First of all, as Semino and Short (2004: 98) themselves note, "speech and writing presentation are … closely related to each other." They both stand in opposition to presentation of thought, which unlike presented speech and writing is "a private phenomenon that is, at best, only partly verbal in nature" (Semino and Short 2004: 98). Speech and writing, on the other hand, "are modes of communication which result in observable and potentially public verbal behaviour." In fact, in my own research, I have used the labels Public and Private discourse that indicated presentation of speech and writing as one category that stood in opposition to presentation of thought.

In addition to speech and writing both being used to communicate, the effects that each of them produces when incorporated into a text are quite similar. Semino and Short (2004: 50) point out that "the writing presentation ... is very like the speech presentation ... in relation to the effects associated with particular categories. This is ... because in both cases the original is ... a piece of discourse, even though the medium is different." Below are the categories of writing presentation (for a comparison with the speech and thought presentation categories see chapter 1). In each case, italicized portions of the examples show instances of writing presentation:

- **Free Direct Writing ([F]DW)**: He wrote in his letter to Mary, *"I miss you very much."*
- **Free Indirect Writing (FIW)**: When I came into the room, I noticed a message scribbled on the wall: *Abandon all you hope*.
- **Indirect Writing (IW)**: In his letter to Mary *he wrote that he missed her very much*.
- **Narrator's Presentation of Writing Acts (NPWA)**: When I came into the room, I saw *him scribbling something on the wall*.

The reason the above examples are not taken from any of the popular science books we are discussing is because presentation of writing in them is extremely rare. It does not come as a surprise though: Semino and Short (2004: 99) also "noted that writing presentation is considerably less frequent ... than either speech or thought presentation."

Besides the rarity of presented writing in popular science, it is sometimes impossible to distinguish between speech and writing presentation, as the authors may present writing in the form of speech by using the scientists' diaries, papers, or even secondary publications as sources for presented discourse, not to mention that often no specific source for a particular instance of discourse is identified. For example, Greene's (2011: 11–12) description of Lemaître's discovery of the Big Bang, shown below, includes several instances of speech presentation from Lemaître and Einstein.

Underlined fragments identify the utterances as speech since they introduce locations and situations where exchange of information is most often achieved orally, and in the case of the last underlined fragment, the combination of the physical action and the reporting verb suggests speech rather than writing:

> "Your mathematics is correct, but your physics is abominable." <u>The 1927 Solvay Conference on Physics</u> was in full swing, and this was Albert Einstein's reaction when the Belgian Georges Lemaitre informed him that the equations of general relativity … entailed a dramatic rewriting of the story of creation…. The universe, Einstein admonished Lemaitre, is not now expanding and never was…. Six years later, in <u>a seminar room</u> at Mount Wilson Observatory in California, Einstein focused intently as Lemaitre laid out a more detailed version of his theory…. When the seminar concluded, <u>Einstein stood up and declared</u> Lemaitre's theory to be "**the most beautiful and satisfactory explanation of creation to which I have ever listened**."

Note that all the indications of speech presentation (rather than writing) come from the author. They either set the scene (conference, a seminar room) or are reporting devices ("stood up and declared"). Presented discourse itself contains no indications of being either speech or writing. Only one instance of direct speech (bolded) is attributed to a source (a secondary one) in the chapter end notes. The reference is insufficient in determining whether the original words were expressed orally or in writing. The shift from the third to the first person narration that the Direct Speech creates changes the perspective of the story but does not help identify this particular instance of presented discourse as either speech or writing. DS in this instance can be easily compared to direct speech in a novel, where the "idea of anterior vs. posterior discourse situations does not sensibly apply at all" since it is impossible to determine what the original was (Short 2012: 20). In my opinion, the difference between speech and writing in this case is not important, and that is probably why a clear distinction is impossible to determine.

The same situation occurs in Kean (2012: 138–141) when he recounts Rous's discovery of a virus that can cause cancer. Speech presentation appears without any references in the chapter, but the end notes indicate that a biography of the scientist was used in creating the story; the reference is not specific enough to identify which instances of the speech presentation it covers, and it is certainly not enough to distinguish between writing and speaking. This makes it problematic to check the potential instances of writing presentation for accuracy—the chief parameter for the division between writing and speech (see Short et al. 2002: 327, Semino and Short 2004: 113). The only undeniable reference to writing in that particular story is a statement that Rous "did publish his results" and a reference to "scientific prose" followed by the following Direct Writing: "It is perhaps not too much to say that [the discovery] points to the existence of a new group of entities which cause in chicken neoplasms [tumors] of diverse character" (cited in Kean 2012: 140). However, even this explicit instance of Direct Writing does not easily lend itself to be checked for accuracy since no title or other identifying information is given for Rous' publication. In this case, and in cases similar to it, presentation of writing is interchangeable with presentation of speech since the type of communicative activity (speech or writing) does not influence interpretation in any way.

For that reason explicit references to writing are not as numerous in popular science books as the instances of speech presentation, and the majority of the reporting verbs are non-specific enough to allow presented discourse to be interpreted as either speech or writing.

Unlike the examples introduced so far, there are definite instances of writing presentation in popular science books that would lose part of their significance if analyzed as speech presentation.

The most common cases of unambiguous writing presentation are connected with mentions of literature. References to literature and to specific works of literary fiction are responsible for the clearer lines between writing and speech. In such cases when a quote from a literary work is incorporated into a popular science text, presented discourse introduces quotations from that literary work, not discourse of individual authors. Consider an example:

> William Shakespeare was born ... in a plague year..., and his career was interrupted several times, when plague epidemics forced the theatres to close down. Shakespeare had Mercutio, in *Romeo and Juliet*, say, "A plague on both your houses!" His audiences would have understood what he meant. Most doctors thought that plague was a new disease, or at least one that Galen had not written about [Bynum 2012: 42].

When Bynum quotes Shakespeare, it is important to let the reader know that he is quoting the play *Romeo and Juliet* and not something Shakespeare possibly said or thought. In other words, presented discourse introduced as the presentation of writing reflects Shakespeare the author, who expresses himself through his characters, as opposed to Shakespeare the person, who would speak/write as himself (in a letter to a friend, for example, if such a thing existed). The same is true for the following examples from Kean (2012: 300–301, 312) as they deal with Tolstoy the author, who is projected through the voice of the narrator in *Anna Karenina*. Kean (2012: 300–301) uses Tolstoy's opening lines of the novel to illustrate a process of chromosome sequencing:

7. Literature and Limericks 119

As observers have noted, the process was analogous to dividing a novel into chapters, then each chapter into sentences. They'd photocopy each sentence and shotgun all the copies into random phrases—"Happy families are all," "are all alike; every unhappy," "every unhappy family is unhappy," and "unhappy in its own way."

Furthermore, context plays an important role in distinguishing presentation of writing from presentation of speech in instances other than those dealing with works of literary fiction. For example, Ferris (1988: 331–332) refers to Howard Georgi "writing a limerick on the blackboard":

> Howard Georgi, known for his work in grand unified theory, introduced a 1984 Weinberg lecture at Harvard by writing a limerick on the blackboard that read:
>
>> Steve Weinberg, returning from Texas,
>> Brings dimensions galore to perplex us.
>> But the extra ones all
>> Are rolled up in a ball
>> So tiny it never affects us.

Carroll (2012: 78) quotes a caption that appeared under "a photograph of the CMS detector next to a photograph of a pigeon" in the *Telegraph*.

> The *Telegraph* printed a photograph of the CMS detector next to a photograph of a pigeon, with the caption, "The Large Hadron Collider (left) and its arch-nemesis (right)."

Carroll (2012: 156) also used Direct Writing when he needed to introduce personal perspective of one of the scientists. The example comes from a story involving Chen Ning Yang, Robert Oppenheimer, and Wolfgang Pauli. Yang was presenting his new theory at a seminar chaired by Oppenheimer, and Pauli kept rudely interrupting Yang. Here is how the story ends:

> The next day, Pauli sent a simple note to Yang: "I regret that you made it almost impossible for me to talk to you after the seminar. All good wishes. Sincerely, W. Pauli" [Carroll 2012: 156].

The narrative makes it clear that Pauli was quite obnoxious during Yang's talk. Carroll (2012: 155) tells the reader, "As an audience member in a scientific seminar, it may occasionally happen that you disagree with something the speaker is saying. The usual protocol is to ask a question, perhaps make a statement to register your disagreement, and then let the speaker continue. That wasn't Pauli's style. He interrupted Yang repeatedly." From this brief excursion into the etiquette of scientific seminars and common knowledge about polite behavior, the reader knows that Pauli behaved discourteously. Yet, his personal message to Yang (delivered in writing) has no trace of remorse or apology. On the contrary, it places the blame on

Yang. Pauli's interpretation of the situation is radically different from that of the author and of the other scientists involved. Using an instance of Direct Writing from Pauli to relay his side is quite effective in presenting Pauli's perspective. The author is using Direct Writing to contrast his own assessment of the situation with that of Pauli.

These and similar occurrences are unmistakably presentations of writing and explicitly intended as such. Introducing them indicates that the authors, on these particular occasions, wish for the reader to process the information as having come from a written source. Mistaking writing for speech would either alter the facts (as in the case of Georgi or Pauli) or would make the discourse nonsensical (in the case of the *Telegraph*'s caption).

Presented writing is usually associated with additional artifacts such as specific forms of writing (e.g., limericks, novels, notes, etc.) or images (as in the *Telegraph* example from Carroll). Once divorced from these contexts, presentation of writing may become unclear to the reader. In a way, what this evidence shows is that writing is a more constrained form of presented discourse as compared with speech. When the contextual background is less important, writing can be substituted for speech—that is what often happens in the narratives of discovery. However, once the background context is vital to the correct interpretation of the message, the presentation of speech cannot be substituted for writing.

This idea is supported in the findings of Short et al. (2002) and Semino and Short (2004: 113), who suppose the presentation of writing more accurate and attentive to the details of the original discourse than presentation of speech. The suggestions of Short et al. (2002) are of particular relevance. Short et al. (2002: 327) argue for a "context-sensitive account of discourse presentation" that gives writing presentation the monopoly on accuracy because of the "checkability" factor. That is, with accurate bibliographical information or enough contextual clues, it is possible to trace a specific instance of writing presentation to its original. This attention to context and the potential for checking an instance of writing presentation for accuracy is connected with what I have been referring to as "artifacts" and "contextual background" of presented writing.

One possible explanation for the lack of references to presented writing in the narratives of discovery and the more numerous occurrences elsewhere in the books is the general broader range of the topics covered by presented discourse when it is introduced outside of discovery stories, which are constricted by their nature. Stepping outside the tales of discoveries also allows the authors to introduce presented voices other than those of scientists more readily. Some of these voices are available only through writing and become less effective as presented speech, as is the case with some literary references. Another possible reason is that outside the narratives of discovery the broader focus allows for introducing contexts other than those of communication between scientists.

Overall, presentation of writing in popular science books functionally mimics presentation of speech expect for those cases when specific contexts play a role in the interpretation of the message. It is important to be aware of presented writing as a category of presented discourse, but it is not always necessary or productive to analyze it on its own terms as it does not always supply any significant additional insight into the use of presented voices in popular science.

8

Definitions
Types and Methods

Writing about scientific advancements, it is impossible to avoid discipline-specific terminology—the jargon that science is famous for and that became one of the key factors in separating amateurs from professionals. The language of professional science did not evolve organically, but was deliberately manipulated into existence by a group of professional scientists with T.H. Huxley and John Tyndall at the head. As Lightman (2000:101) explains, "Huxley and his allies ... worked to purge scientific societies of wealthy, aristocratic amateurs, Anglican clergymen enthralled by natural theology and women with a keen interest in science." Huxley's goal was to establish professional scientists as the only keepers and distributors of Natural knowledge. Now that the doors of the laboratory were tightly shut to all but the select few, Huxley and his like-minded colleagues sought to create a specialized, jargon-laden way of communicating scientific advances that only professionals would be able to handle. "Complicated experiments, perhaps in combination with model building, were described" in such terms that "exclud[ed] the participation of non-professionals." In doing so, this group of self-appointed gate-keepers "dismantled the bridge between elite science and public discourse" (Lightman 2000: 101).

Popular science authors of today are busy putting that bridge back together. However, they cannot deny nor avoid the linguistic legacy of science and therefore have to deal with the jargon-heavy language. One way in which they overcome this obstacle is by introducing a variety of creative and innovating techniques for defining scientific terminology.

Sometimes the authors use traditional methods of defining by genus (the general class the objects belongs to) and difference (particular features that distinguish the object from the others). On other occasions, they use

8. Definitions

figurative language to simplify information or define by action. At times, instead of providing clear definitions of scientific theories or concepts, popular science authors focus on the current state of scientific progress and point out the problems that are still under investigation. These diverse approaches can be combined together in order to produce what I call "definitional strings"—segments of text that explain one term using multiple strategies, addressing the needs of multiple readers with diverse requirements for the depth of explanation. This strategy produces a multi-part definition whose self-contained components can be split throughout the text. Once a reader encountered all of the definition's elements, the full picture emerges, yet a reader is not overloaded with the information all at once. Below is an example of a definitional string for prime numbers from du Sautoy (2011). He starts his definition with

> These are **the primes**, the indivisible numbers that are the building blocks of all other numbers—the hydrogen and oxygen of the world of mathematics. These protagonists at the heart of the story of numbers are like jewels studded through the infinite expanse of numbers [du Sautoy 2011: 6].

He follows up with

> **prime numbers** represent one of the most tantalizing puzzles we have come across in our pursuit of knowledge [du Sautoy 2011: 6].

Later, he provides a less enigmatic definition:

> **A prime number** is a number that is divisible only by itself and 1 [du Sautoy 2011: 7].

And concludes with

> **Prime numbers** are the most important numbers in mathematics because all other whole numbers are built by multiplying primes together [du Sautoy 2011: 7].

This definitional string starts with figurative language that relates the author's excitement and ends with two more substantive definitions that draw on scientific knowledge. The strategy of a definitional string allows the author not only to delimit the term in order to define it ("a number that is divisible only by itself and 1") but also to present enough information to show other facets of the primes—their importance to the field of mathematics ("the most important numbers in mathematics"), the current state of scientific knowledge regarding primes ("the most tantalizing puzzles we have come across"), the correlation with other numbers ("the building blocks of all other numbers"), and the author's attitude ("protagonists at the heart of the story," "jewels").

The definition that emerges from the examination of the string is reader-oriented; it is shaped by the reader's needs as they are perceived by the author (for more on this see chapter 9). This accommodation of the reader's perceived needs by the popular science author and the resultant definitional strings allow me to revise the notion of definition and to suggest less rigid requirements for a definition than the ones proposed by logic and philosophy (see chapter 1 for traditional approaches to definition).

In the remainder of the chapter, I will discuss the individual components of definitional strings and their functions.

Popular science books repeatedly utilize what I call prototypical definitions. These are essentially definitions that state what class the subject belongs to (a piece of equipment, a theory, a person, etc.) and how it differs from other subjects in that category. Such definitions have the A = B structure, where the subject (A) is joined by the hinge (usually expressed in the form of the verb "to be") to the describers (B), which include the class and difference. For example, Albert Einstein (A) is (=) a physicist who came up with the theory of relativity (B). The describers (B) consist of "a physicist"—class and "who came up with the theory of relativity"—difference. There are several variations on the position of the subject (A), the describers (B), and the expression of the hinge (=). See table 8.1 for examples.

Table 8.1. Prototypical Definition

Variations by the position of describers and presence of the hinge	*Representative example*
Subject *hinge* **Describers**	But they had yet to invent **the place-value system**, which *is a way of writing numbers* (class of subject) *so that the position of each digit corresponds to the power of 10 that the digit is counting* (difference) (DuSautoy 2011: 20).
Subject *implied hinge* **Describers**	**Kaluza-Klein theory**, *proposition* (class) *that our universe has spatial dimensions beyond the three of everyday experience* (difference) (Greene 2011:85).
Describers hinge **Subject**	To obtain these uniform magnetic fields, physicists start *with two large coils of wire* (class), *roughly two feet in diameter, stacked on top of each other* (difference). This *is called* a **Helmholtz coil** (Kaku 2011: 61).
Describers implied hinge **Subject**	Chapter 1, "The Curious Incident of the Never-Ending Primes," takes as

Variations by the position of describers and presence of the hinge	Representative example
	its theme *the most basic* (difference) *object of mathematics* (class)*:* **numbers** (DuSautoy 2011: 2).
Describers *hinge* Subject, Describers	*It* (class) *is* **a law of evolution** *that fitter species arise to displace unfit species* (difference) (Kaku 2011: 100).

This group of definitions is the least controversial among the types of definitions found in popular science. Its only unusual feature is the ability to switch the position of the subject and the describers. Some prototypical definitions include an explicit hinge (usually some form of the verb "to be"), and some have an implied hinge. What I call "an implied hinge" is a typographical mark such as a comma or a colon. A hinge in the form of the verb "to be" can be easily inserted in place of the punctuation.

In addition to prototypical definitions, popular science authors also rely on what I labeled "procedural definitions." A procedural definition can be described as a definition by action. Instead of telling the reader what a particular thing is, a procedural definition describes what that thing does. A procedural definition usually functions as a component of a definitional string together with a prototypical definition. If a prototypical definition provides primary information about a term, a procedural definition offers additional details:

> To obtain these uniform magnetic fields, physicists start with **two large coils of wire, roughly two feet in diameter, stacked on top of each other**. This is called a **Helmholtz coil**, [*prototypical definition*] and **provides a uniform magnetic field in the space between the two coils**. [*procedural definition*] [Kaku 2011: 61].

In the above example, "Helmholtz coil" is the subject, and the phrase "two large coils of wire, roughly two feet in diameter, stacked on top of each other" functions as the describers. This is a prototypical definition by class ("coils of wire") and difference ("two large," "roughly two feet in diameter, stacked on top of each other"). What follows is a procedural definition that identifies the subject not by class and difference but by process ("provides a uniform magnetic field"), in which the defined object (Helmholtz coil) is a participant, with the phrase "in the space between the two coils" as the spatial circumstance—an indicator of where the action takes place. The verb "provides" does not only describe the process, but also represents a hinge for the procedural definition. As you can see, in a procedural definition, there is a greater variety of verbs that can function as a hinge. It is, overall, a less formal and rigid way of defining. Unlike in a prototypical definition, the hinge in the example above does not demonstrate a relationship of equality between the subject and the describers. In other words, procedural definitions do not use the A = B structure.

The focus of a procedural definition is on the action that the defined object performs. Thus, in a procedural definition, the hinge is not simply a connecting element between the subject and the describers but an important functional part of the definition, naming the process in which the subject engages. According to Calsamiglia and Van Dijk (2004: 384–385), the process or function of an object described within a definition corresponds to "a general schema of knowledge categories," which means that it adds to our knowledge about that object. Thus a procedural definition is a crucial step on the way to a reader's understanding of scientific terminology and concepts.

I see procedural definitions as a combination of Aristotle's ideas of demonstration and definition. A demonstration (showing what a thing does) reassures us that the defined object really exists, and a definition describes what that object is (see chapter 1 for a discussion of definitions within Logic and Philosophy). This way we can learn about things we have never seen or are unlikely to encounter. Regarded in this way, procedural definition is a linguistic demonstration used when an actual physical demonstration is unavailable. It ties the process of defining to a thought experiment—a tool often used in professional and popular science (see chapter 6).

The example of the Helmholtz coil definition illustrates an instance when the procedural definition follows the prototypical definition. However, procedural definitions may occupy any position within a definitional string. They may also appear on their own without the support of the other types of definitions. Here is an example of a procedural definition functioning independently (the subject is bolded, describers bolded and underlined):

The radiation sent into my body was **non-ionizing** and **<u>could not cause damage to my cells by ripping apart atoms</u>** [Kaku 2011: 59].

This definition represents a trend in popular science texts that Myers (1990) describes as emphasizing of individual objects and their actions instead of describing general processes. Myers (1990: 182) attributes this change in emphasis to "the substitution of popular terms for scientific terms." However, as the definition above demonstrates, it is possible to achieve the same effect preserving scientific terminology. In this case, the author is talking about his own experience with the subject (non-ionizing radiation), thus making the definition even more individualized. Focusing on the object and the process it is involved in creates a definition that is more specific than a prototypical definition of the same term would be *Non-ionizing radiation is radiation that does not rip apart atoms*. By presenting the subject in action, the procedural definition creates a description that is more suitable for a specific popular science text than a generalized prototypical definition would be.

The rationale for including a procedural definition under the broader heading of definition is that it can be easily paraphrased into a prototypical definition, as in the "non-ionizing radiation" example above. In order to derive a prototypical definition from a procedural definition, it is necessary to add a class of the subject since none is provided and then convert the process in the difference using a that-clause:

> Einstein Field Equations ... tell us precisely how space and time will curve as a result of the presence of a given quantity of matter [Greene 2011: 14] [*procedural definition*].
>
> [my paraphrase]Einstein Field Equations are **equations** (class) **that tell us precisely how space and time will curve as a result of the presence of a given quantity of matter** (difference) [*prototypical definition*].

By specifying a class of the subject, the definition becomes more generalized since, according to Aristotle and Plato (relaying Socrates' ideas), a class is designed to supply a general description. However, when the class is absent, the difference (in this case, the process) becomes the primary defining element, and since, as Aristotle and Plato suggest, the difference describes what is particular in an object, the result is a more specific definition.

By their structure, procedural definitions allow the authors to focus on one object instead of defining a whole class of objects. For example, when Kaku (2011: 61) defines a Helmholtz coil using a procedural definition (see example above), he is defining one object and not necessarily all Helmholtz coils. However, it is worth noting that this definition refers to a hypothetical process of stacking coils in order to create a hypothetical object. The author is not referring to a Helmholtz coil in front of him, for example. So while such a procedural definition does not define a class of objects, it is not referring to an actual physical object either. In this case, a procedural definition helps facilitate a thought experiment.

At the same time, procedural definitions can reference specific objects or phenomena. For instance, the procedural definition for "non-ionizing radiation" is clearly focusing on one specific instance when the phenomenon occurred: "I once had my own brain scanned by an fMRI machine.... The radiation sent into my body was non-ionizing and could not cause damage to my cells" (Kaku 2011: 59). Regardless of whether a procedural definition is referencing a specific or a hypothetical object/phenomenon, it can provide the means to focus on one object/phenomenon and not a whole class. Using a procedural definition produces a definition that is especially fitting for a particular text and works well within it but might not be a successful definition outside that context.

The next commonly used type of definitions in popular science I call "delayed definitions." A delayed definition is an incomplete definition that is missing the describers or, less often, the subject. It is used as a technique to get and keep the reader's attention by promising the full definition later in the text. A delayed definition is clearly marked by the author's comments on the absence of a full definition. When using this tactic, the author explicitly avoids producing a complete definitional string. He may, however, provide one component of a string which will present an incomplete definition, as the following example illustrates:

> Intriguingly, there are primes hidden behind these perfect numbers. Each perfect number corresponds to a special sort of prime number called a **Mersenne prime** [*subject*] **(more on this later in the chapter)** [*comment*]. To date, we know of only 47 perfect numbers. The biggest has 25,956,377 digits [du Sautoy 2011: 28].

This example has a clearly identifiable subject "Mersenne prime," and a prototypical definition, "a special sort of prime." However, this definition alone does not provide enough detail and clearly needs supplementation by the other definitions typically found in a string. The phrase "more on this later in the chapter" indicates that the definition is incomplete and more describers will follow. The author's evaluative introduction "Intriguingly" reflects not only his personal fascination with the subject but combined with the delayed definition is designed to create interest in the reader so she reads on. At the same time, the text immediately after the definition demonstrates that the full definition is not vital at this point. The procedural definition comes some 20 pages later. So in a way, the delayed definition could be regarded as a pre-definitional strategy. Other researchers have identified similar occurrences and called them "preliminary definitions" (see, for example, Darian 2003:49).

There are two ways to compose a delayed definition. The author can omit the describers (as in the example above), or he/she can omit the subject itself. The following definition does the latter:

> **The most refined cosmological theories** [*describers*]**, to be encountered shortly** [*comment*], don't lead us anywhere near this possibility [Greene 2011: 36].

8. Definitions

In addition to being an interest-boosting mechanism, a delayed definition lets the author focus the reader's attention on the issue being discussed without providing too much information that he sees as unnecessary at that time. The comment that always accompanies a delayed definition reassures the reader that she is not missing out on important information and will receive it in due course.

Popular science texts often balance on the edge of the present and the future. As Parkinson and Adendorff (2004: 388) suggest, "Popular texts function as narratives of research, reporting on new knowledge claims not yet endorsed as fact by the research community." When dealing with cutting edge scientific advancements and their terminology, the authors may state the term but at the same time indicate to the reader that no definition yet exists. A zero definition is an acknowledgment of the lack of endorsement by the scientific community or, in some cases, the lack of scientific knowledge on the subject. Consider the following example from Greene (2011: 4):

> You'll notice that the terms are somewhat vague. What exactly constitutes a world or a universe? What criteria distinguish realms that are distinct parts of a single universe from those classified as universes of their own? Perhaps someday our understanding of **multiple universes** will mature sufficiently for us to have precise answers to these questions. For now, we'll avoid wrestling with abstract definitions by adopting the approach famously applied by Justice Potter Stewart to define pornography. While the U.S. Supreme Court struggled to delineate a standard, Stewart declared, "I know it when I see it."

This example demonstrates that the phenomenon and the term "multiple universes" are not yet clearly understood nor defined within the scientific community. Attempting to explain multiple universes to the reader, Greene (2011: 4) provides a perfect illustration for a type of definition Harris and Hutton (2007: 11) call "stipulative"—a definition designed for theoretical entities. Below is a stipulative definition:

> With its hegemony diminished, "universe" has given way to other terms that capture the wider canvas on which the totality of reality may be painted. Parallel worlds or parallel universes or multiple universes or alternate universes or the metaverse, megaverse, or multiverse—they're all synonymous and they're all among the words used to embrace not just our universe but a spectrum of others that may be out there [Greene 2011: 4].

It becomes apparent that not only is there no clear definition for what Greene labels "multiple universes," but there is also no one accepted term for the phenomenon. In this case, the function of the definition, once it emerges, will be to "formalize recognition" of one of the terms that Greene names as appropriate to describe the subject (Harris and Hutton 2007: 11).

Zero-definitions, dealing with subjects that are not fully known to the scientific community, inevitably contain metalanguage—that is they frequently comment on the name or names currently used to describe the phenomenon being defined and as such, demonstrate the lack of agreement on the term. For instance, the hesitation about the name and nature of the subject that Greene (2011: 4) exhibits is an indicator of what Verschueren (2004: 61) calls "metapragmatic awareness"—a realization that the meaning and name of the term "multiple universe" is still under investigation by the scientific community. As Verschueren (2004: 69) explains, "metapragmatic awareness ... contributes crucially to the generation and negotiation of meaning," and since meaning is both generated and negotiated through definitions (among other means), the presence of metalanguage is not surprising. Metalanguage would be expected in definitions that are in the process of being constructed.

There are several instances of metalanguage in the above definitions. For instance, when Greene (2011) puts quotation marks around the term ("universe"), he signals that what follows is a *mention* of the term instead of *use*.

The zero-definitions that I introduced above are constructed based on the lack of knowledge by the author and the scientific community he represents. On the other hand, there are cases where zero-definitions are used not because the author does not know the definition, but because he/she chooses not to provide it. Such instances differ from delayed definitions since there is no indication that a definition will be provided anywhere else in the text. Consider an example (subject bolded, describers bolded and underlined):

> Using **<u>something</u>** *called* **a zeta function**, special numbers called imaginary numbers, and a fearsome amoun t of analysis, Riemann worked out the math that controls the fall of these dice [du Sautoy 2011: 52].

In this case, du Sautoy deliberately chooses not to give the reader a definition of a "zeta function," brushing the subject off by defining it as "something." Following the author's lead, the reader is also to disregard the term. However, this example does not represent the majority of the zero-definitions in popular science. Zero-definitions usually include the subject, but rather than providing the describers for it, the authors express the lack of knowledge by the scientific community on the matter, as examples from Greene (2011) demonstrate.

A zero definition stands almost in exact opposition to a procedural definition, which tells the reader what the defined object does and thus (according to Aristotle) confirms the physical existence of such an object. A zero definition tells the reader what is unknown about the object and confirms the fact that scientists have not yet been able to comprehend the phenomenon embodied by the object fully. Sometimes a kind of classification of the potential describers (bolded and italicized) is present:

> To account for the more general possibility that the energy evolves, and to also emphasize that the energy does not give off light (exemplifying why it had for so long evaded detection) astronomers have coined a new term: **dark energy**. "Dark" also describes well the many gaps in our understanding. No one can explain the ***dark energy's origin, fundamental composition, or detailed properties***—issues currently under intense investigation to which we shall return in later chapters [Greene 2011: 24–25].

In this example, the subject (bolded) is clearly identified as "dark energy," and what follows is a list of the potential describers that once understood, will comprise a definition. Interestingly, the presence of the phrase "issues currently under intense investigation to which we shall return in later chapters" is a feature of a delayed definition. Here, however, the comment is designed not to reassure that a definition will come later in the text but to reinforce the lack of definitive knowledge on the subject by the scientific community.

In chapter 5, I talked about metaphors and analogies in presented speech of scientists. Figurative language is a powerful explanatory tool, and it is not limited to presented discourse. There are definitions that rely on figurative expressions as well. I label them "figurative definitions." A figurative definition is a definition that relies on analogies, metaphors and other expressive means and thus embodies the act of imagination on the part of the author and encourages the reader to use his imagination as well. In general, the use of figurative language permeates popular science texts. The authors go as far as to ascribe figurative expressions to scientists when presenting the speech, thoughts, or writing of the members of the scientific community to the reader. In such cases it is debatable who is being creative, the author or the scientist whose words are being paraphrased (for more on this see chapters 5 and 6). In a figurative definition, however, the creativity expressed is undeniably the author's. It is a definition where a writer gets to demonstrate not only his scientific knowledge but also the ability to use language in unexpected ways. Current research (see, for example, Tagg 2009) suggests that creativity involves collaboration between the involved parties and builds on prior knowledge—both aspects are essential to the creation of definitions in general and figurative definitions in particular. In order for a figurative definition to be successful, both the author and the reader need to be familiar with the references used in metaphors and analogies.

The information provided by a figurative definition is not the straight reflection of the knowledge possessed by the scientific community but rather the author's interpretation of this knowledge through the use of an analogy or a metaphor:

> Mathematicians say that the infinite tabletop and the video-game screen are shapes that have constant zero curvature. "Zero" means that were you to examine your reflection on a mirrored tabletop or video-game screen, the image wouldn't suffer any distortion, and as before, "constant" means that regardless of where you examine your reflection, the image looks the same [Greene 2011: 21–22].

This example shows Greene's personal approach to defining "constant zero curvature." Even though the author is not openly asking the reader to imagine anything (there is no directive like "imagine" or "think of"), he calls on the reader's imagination just the same. Using familiar objects (mirror, tabletop) to explain complex terminology is not the only way Greene makes this definition more accessible. Notice that he splits the subject into two parts: "zero" and "constant." In each sub-definition, the subject is clearly identifiable, and so is the hinge "means," which remains the same for both parts of the definition.

Using analogies and metaphors to connect the world of science with the world surrounding the reader is the key feature of a figurative definition and a tradition that goes back to the 19th-century British popularizations. Keene (2014) suggests that it creates "familiar science"—a type of popularization that explains science in terms of the home and its immediate surroundings.

At the same time in modern popular science texts, it is not only the common, everyday objects that function as reference points in figurative definitions. In some instances, basic scientific notions are used alongside everyday objects to create the imagery that becomes a definition. For example, du Sautoy (2011: 53) defines the Riemann hypothesis using the behavior of molecules of gas. He takes for granted that his reader understands what a molecule is because, while explaining the behavior of the particular molecules, he does not define or in any way explain what a molecule itself is. This is a definite shift from "familiar science," which did not expect nor want the reader to possess any prior scientific knowledge, to an explanation that expects the understanding of basic scientific ideas. It is an indication of the more elevated status of the modern reader and of the more respectful and open-minded model of popularization. The expectation that professional scientists read popularized accounts is also a contributing factor (for more on the dual—lay and professional—readership of popular science see the introduction). Here is du Sautoy's example:

> Another way to interpret the Rienmann hypothesis is to compare the prime numbers to molecules of gas in a room. You may not know at any one instance where each molecule

is, but the physics says that the molecules will be fairly evenly distributed around the room. There won't be a concentration of molecules in one corner and a complete vacuum in another. The Riemann hypothesis would have the same implication for the primes. It doesn't really help us to say where each particular prime can be found, but it does guarantee that they are distributed in a fair but random way through the universe of numbers.

Later, du Sautoy (2011: 68) provides a figurative definition that helps the reader understand the structure of a molecule, not the concept of a molecule itself, and he never shows any evidence that he thinks the reader is not at least partially familiar with what a molecule is:

> A molecule can be visualized as a collection of Ping-Pong balls joined together with toothpicks.

Figurative definitions are often exciting and unexpected. They represent the creative side of popular science texts. Table 8.2 provides additional examples of figurative definitions and identifies possible types of the hinge and position of the subject in relation to the describers in such definitions.

Table 8.2. Figurative Definition

Variations	Representative example
Subject *hinge* **Describers**	**A molecule** *can be visualized as* ***a collection of Ping-Pong balls joined together with toothpicks*** (du Sautoy 2011: 68).
Subject *implied hinge* **Describers**	… **the cosmic horizons**—***the patches in the cosmic quilt*** (Greene 2011: 28).
Describers *hinge* **Subject**	***This new way of looking at shapes, in which I'm allowed to push and pull them around as if they were made from rubber or Plasticine***, *is called* **topology** (du Sautoy 2011: 98).

Defining scientific terminology for a non-specialist audience can be a challenging task. In order to appeal to their readers, popular science authors use a variety of definitional techniques which they often combine into definitional strings that include prototypical, procedural, figurative, delayed, and zero definitions. A definitional string may contain all of the above-mentioned types of definition or just two or three. Prototypical, procedural, and figurative definitions are used most often, with a clear preference for prototypical definitions.

The use of definitional strings and unconventional definitions represents a specific pattern of communication between popular science authors and their readers. The definitions used in popular science are reader-oriented. In order to construct effective definitions, writers must try to estimate their readers' cultural backgrounds and prior knowledge. As a result, the authors of popular science texts produce defining mechanisms that venture beyond the common structure of a prototypical definition, which could be expressed as A = B. In popular science texts, the authors have to anticipate the readers' levels of interest and distraction and select what information is vital and what could be omitted.

As I will demonstrate in the next two chapters, being attuned to the needs of a variety of readers is a key feature of popular science texts. The authors approach this task in a number of ways; some of them use traditional interactive means (see chapter 9), and some rely on more unusual tactics, rarely associated with non-fiction (see chapter 10).

9

Interacting with Readers through Definitions

The goal of this and the subsequent chapter is to examine more closely the relationship between the writer and the reader. In this chapter we will focus on such aspects of the relationship as communication, authority, and solidarity. In the following chapter, we will examine the reader as a character in a popular science text. Continuing our discussion of definitions, in the present chapter, I will use definitional strings as examples that illustrate how authors address a variety of readers. I will also discuss the power relationship between authors and readers and analyze the type of solidarity expressed in popular science books through definitions.

In effect, definitions are a response by the authors to the needs of their readers. Bell (1984) introduced the framework of "audience design" to describe a relationship between two communicating parties. His conclusions were drawn from oral interactions; however, as often happens, results achieved through examination of oral communication find support in analyses of written texts as well. What Bell (1984) discovered was that a speaker's style is always affected by her audience. Later research (see, for example, Baumgarten and Probst 2004) demonstrates that the concept can be extended to the discussion of printed texts.

Looking at the definitional strategies is one way to gain insight into the kind of reader the author is responding to in popular science books. Since a popular science writer, similarly to any other "speaker" engaged in mass communication (e.g., TV newscasters, radio announcers), does not have the audience directly in front of him, he has to write to an artificially constructed reader. This reader then becomes both the "addressee"—the person at whom the text is aimed directly—and the "referee"—the person not physically present during the interaction but who is influencing it just the same (Bell 1984). As Bell (1984: 192) explains, it is not uncommon for mass communicative texts to merge the addressee and the referee, "the peculiar nature of mass communication gives the media audience referee like characteristics." The merger of the addressee and referee categories, Bell (1984: 192) argues, creates "ingroup identity" between the writer and the reader that has to be maintained through a certain style of speech or writing. That style inevitably becomes "institutionalized" and "largely predictable," which, in case of print popular science, results in (among other things) a similarity of definition types used.

If definitions are a microcosm of audience design, they demonstrate that the authors are more sensitive to the influence of their readers—"outgroup referees" than to that of colleagues in the same field—"in-group referees." As Bucchi (1998: 2) points out, popular science is read as much by professional scientists as it is by lay readers. Thus, it is not unusual for an author to expect fellow scientists to be among his/her readership. Myers (2003) draws attention to the limited sphere of expertise any given member of the professional scientific community possesses. He demonstrates that popularized accounts may serve as introductions into unfamiliar fields of scientific knowledge and may also function as alert mechanisms that draw attention to particularly significant discoveries of colleagues whose disciplines lie outside of one's professional reach. Thus popular science authors may rightly expect their readers to be fellow scientists, who none the less would welcome some help navigating the material.

The very presence of definitions signals the authors' perceived need to explain terminology to a reader who is unfamiliar with it. In producing popular science books, the authors have to engage in what Bell (1984: 188–189) calls "a long-term outgroup referential shift"—that is, they have to "talk" like lay readers and abandon certain communication conventions of their disciplines.

Later studies incorporating Bell's framework elaborate on the hierarchical roles of the participants involved in a speech act or text construction. Originally, Bell (1984: 159) outlined the following communicative hierarchy: speaker, addressee, auditor, overhearer, and eavesdropper. The speaker and the addressee are the most engaged parties, with the addressee exerting the most influence on the language a speaker uses. The other members of the hierarchy are physically more distant than the addressee and thus exert lesser influence. Ladegaard (1995: 98) suggests that those participants who are regarded by Bell as less influential (due to their physical distance or assigned role [that of auditor or overhearer or referee vs. the addressee]), may exert significant influence on the speaker based on "the power relationship between" the parties involved in an interaction. Thus it would be wise not to discard the potential influence of fellow scientists on the authors. When definitions are concerned, the place where all three participants (the author, the lay reader, and the professional) interface is zero-definition. These segments equate the reader, the author, and the scientific community by acknowledging the lack of knowledge on the part of everyone involved. It is no longer the lay reader alone who does not know what is going on; everybody is in the same situation. Zero-definitions demonstrate what is called reader/writer solidarity.

The influence the reader exerts on the writer is only one side of Bell's audience design. He also points out that the speaker (or the author) is capable of molding the audience (or the reader) just as easily; that is what Bell (1984) calls "initiative shift." By writing in a certain way, an author is not only responding to a potential reader, but is also creating a kind of idealized, imaginary reader that best fits the text. Thompson and Thetela (1995) labeled this hypothetical reader "a reader-in-the-text."

For example, figurative definitions will make sense only to those readers who can decipher the metaphors and analogies of the author. In including figurative language, the author is imagining an idealized reader who is capable of assigning proper meaning to the often modern pop culture-dependent references. Readers who pick up Brian Greene's books, for instance, have to be familiar with old video games, Eric Cartman, and Pringles. Delayed definitions project a different kind of idealized reader—one who will not want to be distracted by a definition at a certain point. Zero-definitions are counting on readers who want to be part of scientific investigations and do not object to the author's admitting lack of knowledge.

Shaping of the readership by the text becomes especially evident when popular science written in English is translated into other languages. As Baumgarten and Probst (2004: 63, 80) and Kranich (2009, 2011, 2016) discovered, German translations of English language popular science texts do not use the vocabulary and syntax common in German but instead recreate the more communicative nature of the English originals. By doing so, the translators are not responding to the readers' demands for more communicative features (the readers are content with the original German popular science that does not exhibit them), but "train" German readers to expect these lexicogrammatical structures in translations from English.

Overall, the relationship between the author and the reader is shaped by both parties—each influencing the other in order to create a text within which both can function harmoniously.

The readership of science (including popular science) is vast and varied (Myers 1997: 43, 2003: 267–269; Hyland 2010: 118). It encompasses people of various cultural and educational backgrounds. The writers are then faced with a task of appealing to the multitude of potential readers and addressing their various concerns and needs. While Hyland (2010: 118) suggests that the diverse readers can be neatly divided into "an elite educated audience" for whom "popular science books are written" and "the public" who get their scientific knowledge from "specialized magazines," these categories are not as clear cut. Popular science books address Hyland's (2010: 118) "public" as well as an "elite educated audience." Defining scientific terminology is one of the ways writers recognize and construct the background knowledge of the potential readers. As Hyland (2010: 121) notes, "Popularisations ... have to make connections to what readers are likely to already know. This involves constantly defining new concepts."

The strategy of a definitional string supports the idea that the authors are targeting audiences with different educational backgrounds in an attempt to address the needs of a wide range of readers. Different types of definitions that a string contains present scientific terminology from a variety of angles—each definition potentially addressing different needs that different readers might have. For example, here is a definitional string for the "four-sided tetrahedron" from Marcus du Sautoy (2011: 67):

> The four-sided tetrahedron, or triangular-based pyramid *(prototypical definition)* encloses the least volume for a given surface area *(procedural definition)* it is the shape that requires the fewest number of faces to make *(prototypical definition)*
>
> Consider how a potato-chip bag is made. A cylindrical tube is sealed at the bottom, filled with chips, and then sealed in the same direction along the top. But look what happens if instead of sealing the top in the same direction as the seal at the bottom, you twist the bag 90 degrees and then seal it. Suddenly, you're holding a tetrahedral bag in your hand. The tetrahedron has six edges: two where the seals have been made, and four that link the two seals—an edge runs from the end of each seal to each end of the opposite seal *(figurative definition using analogy)*

The first component of the string might be an adequate definition for a reader who can easily visualize a "triangular-based pyramid." On the other hand, the last portion of the definitional string targets the reader who might have trouble imagining such an object. The definitions in the middle might be effective for a more detail-oriented reader; however, they function best within a string and not on their own. If all the definitions in the string are regarded as supplementing each other and functioning as parts of a whole, the result is a definition that is precise and detailed yet relates the term to the everyday reality of most readers.

It has been well established that popular science texts are reader-oriented (see, for example, Myers 2003, Baumgarten and Probst 2004, Liao 2010, Hyland 2010). Perhaps the reason for such an approach is determined by one of the goals of popular science—"to arouse the interest of readers and involve more lay people in the world of science" (Liao 2010: 45). Popular science, according to Liao (2010: 45), possesses "salient features of interaction between writers and readers." Liao (2010), Hyland (2005a, 2005b), and Varttalla (1999) see interaction as recognizing the reader and encouraging him/her to perform certain actions. For instance, Hyland (2005b: 365) lists such elements as "direct reader references" (pronoun "you"), "references to shared knowledge" (the shape of a Pringles chip, for example), and "asides" (short commentary from the author) together with "directives" and "questions" to form one category of linguistic features that "involve the reader in the communication process."

Hyland (2005a: 37) sees "interaction" as an umbrella term that includes elements that acknowledge the presence of the reader and call the reader to action. At the same time, in certain instances, it is best to separate interactive (or reader engagement) elements into two distinct groups: one dealing with recognizing the presence of the reader in a text (e.g., reader pronouns or references to shared knowledge) and the other with asking the reader to do something (directives and questions). I refer to the first group of reader engagement elements as "reader recognition elements" and to the second one as "reader action elements."

The difference between reader recognition and reader action elements leads to two different views of the popular science readership. The first presumes a passive reader who is content with being talked to, and the second presupposes an active reader who would like to be engaged with the text and perhaps with the scientific world (For a discussion of active and passive readers of popular science see Topham 2000, Mellor 2003, Myers 2003). Both reader recognition and reader action elements are important for the success of popular science; however, they perform slightly different functions and are not interchangeable. Definitional strings contain both action elements and reader recognition features, and this variety of reader engagement mechanisms makes them especially effective at addressing the needs of popular science's diverse readership.

Reader recognition and reader action markers are present in different parts of definitional strings. Let's look again at the definitional string for the four-sided tetrahedron from du Sautoy (2011: 67):

> The four-sided tetrahedron, or triangular-based pyramid *(prototypical definition by class [pyramid] and difference [triangular-based])* encloses the least volume for a given surface area *(procedural definition)* it is the shape that requires the fewest number of faces to

make *(prototypical definition by class [shape] and difference [that requires the fewest number of faces])*

Consider how a potato-chip bag is made. A cylindrical tube is sealed at the bottom, filled with chips, and then sealed in the same direction along the top. But look what happens if instead of sealing the top in the same direction as the seal at the bottom, you twist the bag 90 degrees and then seal it. Suddenly, you're holding a tetrahedral bag in your hand. The tetrahedron has six edges: two where the seals have been made, and four that link the two seals—an edge runs from the end of each seal to each end of the opposite seal *(figurative definition using analogy)*

The string concludes with a figurative definition, which contains directives ("consider" and "look") in addition to reader pronouns ("you"). With the two directives, the definition encourages the reader to imagine the process of creating a tetrahedron and presents this process as an easy one to duplicate if the reader is interested in a hands-on experience. On the other hand, the remaining definitions in the string are not interactive in the same sense, and the only reader recognition markers are non-verbal, the typographical markers (commas) that set off the describers in the prototypical definition. Including both reader recognition and reader action elements in a definitional string allows the authors to extend scaffolding mechanisms for the potential readers. If a definition with reader recognition elements does not explain the information sufficiently, the one using action markers might provide enough direction to help the reader understand the definition by constructing a mental image. While directives are the most popular action markers present in definitions, reader recognition elements range from non-verbal markers and reader pronouns to attitudinal markers, which I consider some of the most interesting.

Definitions in popular science books continuously draw on writers' emotions and attitudes to introduce scientific terminology as exciting and memorable. While I regard attitudinal markers as reader recognition elements, Martin and White's (2005: 95) analysis suggests that attitudinal markers are potentially calls to actions since "when speakers/writers announce their own attitudinal positions they not only self-expressively 'speak their own mind,' but simultaneously invite others to endorse and to share with them the feelings, tastes or normative assessments they are announcing." However, even in this view, attitudinal markers are only covertly action-oriented. I find the reader recognition properties of attitudinal markers more important because they indicate that the writers anticipate possible reader responses and react to them. As Martin and White (2005: 45) point out, attitudinal markers indicate writer anticipation "that a given proposition will be problematic ... for the putative reader, or ... that the reader may need to be won over to a particular viewpoint."

I should point out that Martin and White (2005) in arguing this position deal with persuasive texts; however, it is possible to extend their argument to include popular science texts as well, especially since popular science books display a degree of persuasiveness when in comes to certain issues. For example, Greene (2011) is a strong proponent of string theory and defines it as an "approach to unifying all of nature's laws" despite the fact that "string theory has yet to make definitive predictions whose experimental investigation could prove the theory right or wrong" (Greene 2011: 7, 72). Other scientists and popular science authors, Lee Smolin (2007), for example, who do not share Greene's excitement about string theory, prefer to dwell on Greene's second statement and claim that physics has made little progress since the 1970s. So when Greene (2011) presents string theory as a theory that "provided a long sought quantum theory of gravity," he is engaging in essentially the same activity as the authors of the persuasive texts that Martin and White (2005) analyze.

The authors of popular science clearly anticipate the difficulties the readers might have understanding scientific terminology and accepting it as an important and necessary part of the text. This is why the writers' express their positive attitudes in definitions. For instance, Kaku's (2011: 126) definitional string for "tissue engineering" contains a strong positive attitudinal marker:

"Tissue engineering" is one of the hottest fields in medicine, making possible a "human body shop."

This is a prototypical definition that appears at the head of the string and is designed to showcase the importance and relevance of this medical field using informal vocabulary that will appeal to a certain reader. It clearly marks the writer's attitude as positive. This kind of attitudinal marker is an effective introductory device that potentially creates curiosity in the reader. However, this first definition does not tell the reader what exactly "tissue engineering" is. In this case, it is the procedural definition, introduced later on the same page, that is more informative:

Tissue engineering grows new organs by first extracting a few cells from your body. These cells are then injected into a plastic mold that looks like a sponge shaped in the form of the organ in question. The plastic mold is made of biodegradable polyglycolic acid. The cells are treated with certain growth factors to stimulate cell growth, causing them to grow into the mold. Eventually, the mold disintegrates, leaving behind a perfect organ [Kaku 2011: 126].

From this definition the reader learns what exactly tissue engineering involves. The combination of the prototypical and the procedural definitions is designed to work together. The first definition raises the interest and assures the reader that the term at hand is important. Only after this introduction is the reader given the technical details, which at this point, he/she should receive with more attention and enthusiasm.

Another way to enthuse the reader when it comes to scientific terminology is to use figurative language in definitions. Figurative language, as we have already examined, serves an important role in expression of creativity. According to Tagg (2009: 163), creativity and reader recognition go hand in hand. Tagg (2009: 160) describes creativity as "collaborative" and "building on existing knowledge" of the parties involved. She also points out that figurative language functions to signal "involvement," "convergence between speakers," and "a sense of belonging" (Tagg 2009: 163). Such approach to creativity explains why popular science authors choose to embed attitudinal markers in analogies and metaphors. They seize the communicative aspect of the creative process to project their own positive attitudes to the reader. For example, when du Sautoy (2011: 6) defines prime numbers as "the hydrogen and oxygen of the world of mathematics," the metaphor aims not so much at the creation of a memorable image but at reinforcing the author's attitude that prime numbers are very important. Declaring the primes as essential as water is likely to spike readers' interest; at the same time, this figurative definition is relating "the strange and exotic" (the scientific term "prime numbers") to "the commonplace and unexceptional" (Hyland 2010: 121)—a typical trajectory for a definition of any kind.

The primary goal of figurative definitions like the one of prime numbers by du Sautoy is to make the reader pay attention to the term and to transmit the author's obvious interest in the subject—in other words—these kinds of definitions communicate the information through the affective dimension by employing attitudinal markers rather than presenting the information from a purely cognitive point of view, as a traditional definition would. According to Ainley, Corrigan, and Richardson (2005), affect is extremely important in determining whether or not readers choose to engage further with a popular science text. They propose that the emotions of interest and surprise strongly correlate with a reader's decision to continue reading. Ainley et al. (2005: 440) report, "An activating emotion such as a surprise, with its implication that something unexpected or novel has been encountered, would ... show higher intensity associated with the decision to continue." While these findings do not explain what particular features of the texts they examined triggered interest or surprise, my analysis of figurative language and attitudinal markers points to a strong possibility that these textual elements contribute to the creation of an emotional response from the reader. In the very least, this is likely the intention of popular science authors.

Figurative definitions draw on the powers of expressive language. However, not all instances of figurative definitions employing metaphors or analogies contain attitudinal markers. The following definition does not reveal the author's attitude toward the terminology being defined:

> Mathematicians call the cue ball's surface a two-dimensional sphere and say it has **constant positive curvature**. Loosely speaking, "positive" means that were you to view your reflection on a spherical mirror it would bloat outward, while "constant" means that regardless of where on the sphere your reflection is, the distortion appears the same [Greene 2011: 21].

Even when they do not contain attitudinal markers, metaphors and analogies can be seen as reader recognition strategies since they "predict and respond to possible objections" (Hyland 2005b: 366) readers might have to the use of discipline specific terminology. Using figurative language allows the authors to entice and not bore the readers while explaining the concepts unfamiliar to them. For example, when Greene (2011: 22) uses a Pringles potato chip to define "constant negative curvature," he simplifies the language involved in the definition and creates a connection between an abstract scientific term and an everyday object:

> **constant negative curvature** ... means that if you view your reflection at any spot on a mirrored Pringles chip, the image will appear shrunken inward.

While such a definition does not indicate Greene's attitude towards "constant negative curvature," it anticipates the difficulty the reader might have with the term if it were explained in a more traditional way. In this case, the author resolves this difficulty not by activating an emotional response from the reader but by helping him/her create a memorable visual image which embodies the definition. According to Tlauka and McKenna (1998), mental images can be as effective as pictorial images at enhancing understanding: "The cognitive processes mediating imagined stimuli and responses are functionally equivalent to those mediating physical (perceptual) stimuli and responses" (69).

Thus figurative language serves a dual purpose when employed within definitions: it is used to indicate the authors' attitude, activating the reader's emotions as well, and to create memorable visual images. Within figurative definitions, metaphors and analogies act as reader recognition elements as they anticipate and address problems some readers might have understanding scientific terminology. At the same time, figurative definitions evoking emotions are prime examples of the authors shaping their readers. Considering

Martin and White's (2005) argument that when an author demonstrates his/her attitudes towards something, this author is trying to persuade the reader to feel the same way, it is possible to suggest that certain figurative definitions are designed to influence the reader in such a way that his/her attitudes towards specific terminology become the same as the author's. This also happens with presented discourse, as we have seen in chapter 5. In this light, figurative definitions are not simply a response to an audience but also an attempt to shape and control it.

Another noteworthy interactive strategy is a delayed definition. Its sole purpose is to tell the reader that the definition will not be offered at this time and possibly direct him/her to a place in the text where the definition is provided. However, the delayed definitions I encountered rarely contain any explicit reader recognition or reader action markers.

Delayed definitions can be regarded as announcements, metadiscourse features "which look forward" (Hyland 2005a: 34). It is also possible to describe a delayed definition as a proleptical narrative device (Toolan 2001: 43) that gives the reader a glimpse of what will follow. It is possible to interpret delayed definitions as either reader recognition devices or as calls to action. I tend to see delayed definitions as primarily reader recognition devices since they clearly indicate to the reader that the full definition will not be provided and do not explicitly prompt any specific behavior from the reader. Consider an example:

> This is a perfectly sensible geometrical space called a two-dimensional torus. I discuss this shape more fully in the notes [Greene 2011: 21].

On the other hand, it is possible to argue that the indication of where a definition could be found ("I discuss this shape more fully in the notes" [Greene 2011: 21] or "which we shall explain in more detail in Chapter 4" [Kaku 2011: 62]) is a call to action. A reader can, potentially, upon encountering such a statement, turn to the notes or to the mentioned chapter; however, while this opportunity is available, the action is not essential to the understanding of the text. If a delayed definition is a call to action, it is not so for all readers, as some of them will be fully satisfied with not needing to know the definition in order to continue reading and understanding the text.

Writers employ delayed definitions to ease comprehension. In fact, delayed definitions reassure the reader that the definition is not essential at that point. Thus reader recognition properties of delayed definitions are more prominent than any potential reader action features. Delayed definitions reduce the immediate cognitive load. Each time a definition is delayed, the reader is not missing essential information but avoids over-saturation. Let's look at another example:

> Intriguingly, there are primes hidden behind these perfect numbers. Each perfect number corresponds to a special sort of prime number called a Mersenne prime (more on this later in the chapter). To date, we know of only 47 perfect numbers. The biggest has 25,956,377 digits. Perfect numbers that are even are always of the form 2^{n-1} ($2^n -1$) [du Sautoy 2011: 28].

This delayed definition of Mersenne prime allows the reader to focus on the general idea instead of getting distracted by a definition which will not enhance the author's main point. In fact, delayed definitions help the authors create the kind of reader they need—the reader who does not know the definition just yet. The above segment is perfectly clear without the definition for Mersenne prime present. In a way, a delayed definition is a focusing device that allows the author to guide the reader through the text with minimal distractions and the maximum of useful information. Viewed from this perspective, the potential reader action feature of delayed definitions is counterproductive. However, if a reader chooses to find the definition, he/she can do so using selective scanning (Macedo-Rouet et al. 2003: 123)—a reading strategy that allows the reader to skip parts of a text in order to focus on the material he/she finds more important.

In summary, delayed definitions are an example of the authors' willingness and ability to approach various types of readers with different informational requirements and cognitive capabilities. While their primary function is reader recognition (anticipation of confusion and/or distraction), a delayed definition can be regarded as an author's attempt to shape the reader by telling him or her that some information exists but should not be accessed just yet. A reader ignorant of that particular piece of information is the desired reader at that point in the text. Delayed definitions, on the other hand, are not the only means by which the authors demonstrate their control of the material and of their readers.

It is often a given that authors of popular science books are experts. They may be practicing scientists (e.g., Brian Greene and Michio Kaku) or experienced science journalists (e.g., Carl Zimmer, Natalie Angier, or Bill Bryson). In either case, they know a lot more about science (or at least the branch of science they are covering) than their readers do. This is why the readers trust them as guides through the world of science and through the text of the book.

However, even with their expert status established and known to the reader in advance, the writers use specific language that further underlines their authority. For instance, popular science authors use what Gűlich

(2003: 254) calls "self-categorization." Such expressions as "we mathematicians" (du Sautoy 2011: 98, 82), for example, categorize the author as part of an expert group that excludes the reader. Other, less explicit, ways of establishing expert status include references to one's professional work ("When I started working on string theory" [Greene 2011: 91] or "I had a chance to witness these technologies first hand when I visited the CAVE ... at Rowan University ... for the Science Channel" [Kaku 2011: 34]). According to Gűlich (2003: 254), these writers put up such displays because "Experts are not just experts through having acquired certain competences ... but mainly because they present themselves as such in communication."

While Gűlich's (2003) analysis is of spoken expert-to-lay-person interaction, the general aspects of her findings are relevant for written communication as well. For example, Gűlich (2003: 254) suggests that the very act of "explicit introduction of specialist terms" (use of definitions) signals the expert status of the authors. One key aspect of interaction between experts and non-experts that Gűlich (2003: 254) noted in the face-to-face sessions and that is absent from popular science books is what she called "other-categorization." That is, the non-expert audience is given a specific label that designates its lower status: for instance, "patients" as opposed to "doctors." In popular science books, the authors do not use category labels such as "non-experts" or "laymen" to address their readers. For example, you will not find sentences like this one: "For us physicists this is a bigger problem than for you laymen." Reader recognition is established through more power-neutral means, some which we have already covered and more of which will be introduced in chapter 10. One reason for the lack of such explicit power status markers is the diversity of the audience for popular science, which often includes experts. Another reason is the general focus on the reader and the desire to present science and the people associated with the scientific community in a positive way.

Following the terminology of the audience design framework (Bell 1984), I propose that there is a two-fold power structure within popular science books: *initiating power structure*, where the author dominates in light of his expertise and *responsive power structure*, where the author reacts to the readers' needs. Definitional strings can combine both approaches because individual definitions could represent either initiating power structure or responsive one. For example, figurative language of some definitions indicates a responsive power arrangement: the author anticipates a difficulty with a technical explanation and offers a more creative way to understand a term or a concept. On the other hand, references to scientific knowledge (using the phrase "hydrogen and oxygen" instead of the word "water") shows an initiating power structure: the author uses basic scientific facts that he expects the reader to be familiar with, as a scientist would.

Bell (1984: 186) argues that the very "essence of initiative style shift [initiating power structure—in this case] is to address persons as if they were someone else."

In the definition of the Reinmann hypothesis, du Sautoy (2011: 53) assumes that a reader knows how "molecules of gas in a room" behave. His whole explanation hinges on that piece of knowledge. By implying that the reader has solid knowledge of molecular structure and behavior, du Sautoy flatters the reader and, at the same time, projects the kind of reader he wants for his text.

Flattery is not uncommon when initiating power structure is at play. For example, Vartalla (1999:193) demonstrates that a result similar to du Sautoy's (2011: 53) could be achieved through the use of hedges that "enhance the readership's self-image." That is, an author may express doubt about a proposition or not fully endorse a proposition with the idea that the reader will be able to understand his/her hesitation. Hedges (words like "perhaps," "possibly," "appear," "seem," etc.) are common in professional scientific discourse. It was a common assumption that popular science lacks such linguistic devices. However, as Vartalla's (1999) and Kranich's (2009, 2011, 2016) studies show, popular science embraces hedging but uses it in a different way from professional research articles. As Vartalla (1999: 193) explains, "By hedging popularizations writers can not only make their accounts correspond to the preconceptions of the audience, but also enhance the readership's self-image by emphasizing closeness between the author and the reader by claiming common ground in the form of expressions clearly associated with specialist-to-specialist communication."

The effect of sudden closeness and the claims of the common ground are possible only if the initial relationship between the reader and the author implies power imbalance. So while it may appear that flattering the reader is an example of equal footing, it is not so.

More extreme examples of initiating power structure where the reader, the writer, and the scientific community are equal could be found in zero-definitions. In these segments, however, the author is not extending his power to include the reader, but is rather as powerless as the reader, as is the scientific community. By admitting to not knowing everything, the scientific community appears less intimidating and more relatable, and while some readers will see this as an ultimate expression of solidarity, let's not forget that not knowing the exact make up of dark energy is not the same as not knowing what a molecule is. My point is to draw your attention to the artificiality of this power arrangement. Keep this in mind as you read the following section.

There is a view that in popular science, the relationship between the writer and the reader is similar to that of equals, with the writer and the reader presenting a united front in the face of the scientific community. According to Parkinson and Adendorff (2004: 389), "popular articles show solidarity with the reader … by distancing the reader and writer from scientists as a group." Parkinson and Adendorff (2004) base their claim on the analysis of popular articles written by non-scientists. However, the underlying idea that the writer does not always present him/herself as more powerful than the reader is valid even for popular science books written by scientists.

This shift in the power structure between the writer and the reader is evident in zero-definitions, where the author openly admits the lack of knowledge not only on his/her part but on the part of the scientific community as a whole. In this case, the author and the reader are on even grounds, but they are not juxtaposed against the scientific community, as is the case with the texts Parkinson and Adendorff (2004) analyze. The kind of reader solidarity discussed by Parkinson and Adendorff (2004) is *exclusive* with respect to the scientific community. It would be exemplified by such phrases as "what physicists call…" used by both Kaku (2011) and Greene (2011) in order to establish unity with the reader against the scientific community. Even though both authors are physicists, to establish solidarity they choose not to present themselves as such.

A zero-definition, on the other hand, creates *inclusive* solidarity. It is solidarity on a greater scale. The reader's knowledge (or lack thereof) is equated not only with the author's but with that of the scientific community. Thus zero-definitions are one of the possible locations for inclusive solidarity. This definitional strategy lifts the curtain on the scientific issues currently under consideration and potentially invites the reader to join in the effort to solve the puzzles. The communicative aspect of zero-definitions corresponds directly with the goal of popular science to "involve more lay people in the world of science" (Liao 2010: 45).

Zero-definitions, however, are not the only possible places where such solidarity occurs. For example, various asides and remarks directed at the reader can have a similar effect:

> We don't yet know whether there can be odd perfect numbers [du Sautoy 2011: 28].

The pronoun "we" in this case refers not only to the reader and the author but to mathematicians in general as well, neither of whom knows the answer.

I started this chapter saying that the readership of popular science is diverse and the authors need to address a variety of potential readers. It is not surprising that the texts that need to engage diverse readers are targeting them through multiple strategies; what is more unusual is that these strategies manifest in definitions—text segments that are for the most part regarded as informative, not communicative. When Aristotle or Socrates talk about definitions, they are concerned with the construction of knowledge not with the best way to relate this knowledge to someone else. When Robinson (1962: 2–3) lists 13 different statements on what a definition is, they all focus on knowledge (the main idea, essence, etc.) of the defined object. However, when I look at the kinds of definitions found in popular science, I see that they offer much more than bare information. They use figurative language, showcase and evoke emotions, and play with the reader's self-image through the use of initiating power structure. If "there is no single public for science" (Myers 1997: 43), there is also no single way to define scientific terminology in popular science books. Popular science authors take full advantage of the linguistic techniques available to them and reinvent definitions as text segments that are interactive and engaging.

10

The Fictionalized Reader

Before I start talking about the main subject of this chapter—a reader of popular science as a fictional character in the texts—it will be useful to take one more look at fictionality. There are many definitions of fictionality ranging from the traditional approaches that suggest representation of imaginary scenarios and characters (see, for example, Fludernik 1996) to more extreme views that argue any hypothetical scenario is fiction (Skov Nielsen et al. 2015a,b). Some researchers (see, for example, Cohn 1990, Short 2007 and Dawson 2015) propose that inclusion of thought presentation automatically signals fictionality since it is only in fiction that a reader *knows* what other people are thinking and because it is only in fiction that thoughts are presented as well-constructed linguistic units. However, as I have demonstrated in chapter 6, thought presentation, especially in popular science, is projecting not so much the inner world of the scientists but serves as one more communicative tool for the discussion of scientific findings.

The first approach to fictionality is also known as the literary kind of fictionality because this is what we usually come across in literary fictions (novels, short stories, narrative poems). It is the most recognized manifestation of fictionality. Fludernik's (1996) definition of fictionality exemplifies this approach: literary fictionality is understood as "the subjective experience of imaginary human beings in an imaginary human space" (Fludernik 1996: 39). Fludernik (1996) emphasizes "the subjective experience," saying that literary fictions "set out to represent" the human experience and create an "evocation of 'real-life' experience" (Fludernik 1996: 41, 12). To be considered fiction, a text must project the emotional experiences of its characters since emotional reactions are more unique and subjective than physical responses. Dramatization not only of external events but also of inner states is therefore essential to literary fiction.

It appears unreasonable to expect this kind of fictionality in a popular science text, yet this is exactly what popularizers do in order to make their information more relatable to their audiences. I should note at this point, that this is not the only way fictionality manifests in popular science books. The subject is vast, and the information on all aspects of fictionality in popular science can fill its own separate book (I return to it briefly in the conclusion); here, however, we will focus on just one manifestation to continue with the topic of reader and writer relationships.

The idea that a reader exists not only as a concrete person who is holding a book in his or her hand but also as a presence that lives inside the text began to be explored fully in the 1970s with the work of Wolfgang Iser (1972) titled *The Implied Reader* (published in English in 1974). Jonathan Culler (1982) continued the exploration of the implied reader in fiction. Later, Norman Fairclough (1989) and Mary Talbot (1995) looked at the notion of a generalized reader in non-fiction. However, the most popular and complete exploration of this idea comes from Geoff Thompson (see Thompson and Thetela 1995, Thompson 2001, 2012). He is the originator of the now widely-used label the "reader-in-the-text."

Thompson (2012: 80) suggests that his "admittedly clumsy formulation" has "the advantage" of focusing on the "evidence in the text itself." However, his reader-in-the-text, despite concrete evidence of his existence, remains mostly an ethereal entity that, in Thompson's (2012: 80) words "haunts all discourse" and whose opinions and preferences cannot be expressed directly but only through the voice of the author. In other words, the presence of the reader could be discovered by scholars who, not unlike archeologists, have to carefully peel away layer upon layer of textual "debris" in order to reach the precious artifact. If you are curious, some of these scholars are Thompson (2001), Martin and White (2005) and Lewin and Perpignan (2012). In addition to the general elusiveness of the reader-in-the-text to the untrained eye of the non-professional, Thompson (2012: 80) points out that it is "unpredictable" how a real-world reader will fit "into the semiotic shape moulded ... by the text." That is, it's impossible to know if the actual reader will hold the same ideas and attitudes as the ones expressed by the reader-in-the-text. After all, this kind of reader manifestation does not have her own voice and has to rely on evaluation (expression of opinion introduced by the author) and other textual mechanisms (see Thompson 2012: 81) that deal with covert representation of the reader.

The authors of popular science offer a solution to such indeterminacy by introducing a more direct approach to the incorporation of the reader, the technique I call the "fictionalized reader." This kind of reader is created through his or her own presented discourse and functions like any other character in a text. Authors of popular science employ hypothetical Direct Speech to enact possible reader reactions to explanations provided in their texts. Quite often this hypothetical Direct Speech of the reader forms dialogues either with the author or with other characters in the books, who are, like the reader, given voice by means of presented discourse. The reader speaks for himself. There is no longer any need for careful scholarly analysis to determine what the reader thinks or expresses. It's obvious. Consider an example:

> Imagine it's a hot summer night and there's an annoying fly buzzing around your bedroom. You've tried the swatter, you've tried the nasty spray. Nothing worked. In desperation, you try reason. "This is a big bedroom," you tell the fly. "There are so many other places you could be. There's no reason to keep buzzing around my ear." "Really?" the fly slyly counters [Greene 2011: 29].

By making the reader effectively a character in the story, the author eliminated Thompson's (2012: 80) concern for unpredictability of the real reader's reaction. The author is not dealing with a real-world reader any longer but has constructed a thoroughly fictional character. Unlike presented discourse of scientists, presented discourse of a fictionalized reader is not based on any real-world utterances or writings and is, therefore, closer to presented discourse found in fiction.

The fictionalized reader also eliminates the need to have the author as a representative of the reader. Now there is a character with his/her own presented voice who fulfills this function. The author, thus, is free to speak *to* the reader without speaking *for* him/her also. This appears to be a significant benefit for popular science because the fictionalized reader shows up almost exclusively in thought experiments—segments that are heavy with scientific information—where the author presumably needs to concentrate his/her explanatory abilities and not strain the resources by anticipating possible reactions or incorporating possible values of the reader. The fictionalized reader is a perfect mechanism for supplying a reader-oriented interpretation of a difficult issue. Thought experiments that involve the reader are parts of popular science texts that attempt to translate abstract ideas into concrete examples through the application of the scientific principle being discussed to everyday situations. Thought experiments are often introduced in the form of games, narratives, or hypothetical scenarios that feature the reader as the main participant.

The fictionalized reader in addition to simplifying the writer's task by supplying concrete reactions and eliminating guess work also helps the real reader. As Thompson (2001, 2012) mentions, if a real-world reader is not in agreement with the attitudes and positions of the reader-in-the-text, the communication breaks down. However, the technique of the fictionalized reader allows the real reader to step back and experience explanations and arguments through another character in a text without breaking the communication line with the author even if the reader does not see herself behaving or reacting exactly as the fictionalized reader.

The variety of presented discourse types that are assigned to the fictionalized reader suggest that this is a fully functional and well-developed phenomenon. As examples in Table 10.1 demonstrate (bolded text segments represent speech and thought acts attributed to the fictionalized reader), the variety of presented discourse types assigned to the fictionalized reader is comparable with the kinds of discourse presentation assigned to scientists (see chapters 5 and 6). It is also comparable with the types of discourse presentation of characters in novels as analyzed by Semino and Short (2004). For instance, both presentation of speech and presentation of thought are present.

Table 10.1. Discourse Presentation Types Used to Create the Fictionalized Reader

Type of Presented Discourse	Example
Direct Speech	Imagine you work for the notorious film producer Harvey W. Einstein, who has asked you to put a casting call for the lead in his new indie, *Pulp Friction*. **"How tall do you want him?"** you ask. **"I dunno. Taller than a meter, less than two"** (Greene 2011: 152).
Narrator's Presentation of Speech Acts	... **you ... tell the robotic cook in your kitchen to make breakfast and brew some coffee, and order your magnetic car to leave the garage and be ready to pick you up** (Kaku 2011: 354).
Free Direct Thought	Now you face a decision. How many actors should you have at the audition? You reason: **If W. measures heights to a centimeter's accuracy, there are a hundred different possibilities between one and two meters** (Greene 2011: 152).
Direct Thought	Your first thought is, **"Well, protons smash together, the Higgs comes out"** (Carroll 2012: 166).

Type of Presented Discourse	Example
Narrator's Presentation of Thought Acts	Curled up under a warm duvet, just regaining consciousness but not yet having opened your eyes, **you'll remember the Zaxtarian deal. At first it will seem like an unusually vivid nightmare, but as your head starts to pound you'll recognize that it is real**... (Greene 2011: 230).

The inclusion of the reader in thought experiments demonstrates that he/she is able to contribute to the discussion and application of scientific concepts. Let's return to the conversation between the reader and the fly introduced in Greene (2011: 29):

> "How many places are there?" [the fly asks] In a classical universe, the answer is "Infinitely many." As you tell the fly, he ... could move 3 meters to the left, or 2.5 meters to the right, or 2.236 meters up.... Since the fly's position can vary continuously, there are infinitely many places it can be. In fact, as you explain all this to the fly, you realize that not only does position present the fly with infinite variety, but so does velocity.

The fictional reader-character takes on the task of explanation. It is the fictionalized reader alongside the author who educates the fly: "As *you* tell the fly," "as *you* explain all this to the fly." In Greene's thought experiment, the fictionalized reader is also capable of drawing scientific conclusions and realizations, "*you* realize that." The reader pronoun in this case refers to the reader-character not to the actual person holding the book in his hand.

Here is another example. In this scenario the author is teaching the fictionalized reader how to win at the game of nim using mathematical analysis:

> At some point, one of you will remove bars of chocolate so that the piles have 000,000, and 000 in them. *Who does that?* Your opponent always leaves an odd number of 1s in at least one of the piles, so it must be you that makes this move [du Sautoy 2011: 139].

The fictionalized reader is introduced through Free Direct Speech (Direct Speech version of the same question would be "'Who does that?'"). The question is the reaction to the events of the hypothetical game. It is an acknowledgment of the reader's involvement in the thought experiment.

While most often appearing in thought experiments, the fictionalized readers can have their own narratives where they are the primary characters, who interact with other characters via a variety of presented discourse forms. One of the best examples of this, in my opinion, is the concluding chapter of Kaku's (2011) book *The Physics of the Future*. It begins in the following way:

> After a night of heavy partying on New Year's Eve, you are sound asleep. Suddenly, your wall screen lights up. It's Molly, the software program you bought recently. Molly announces cheerily, "John, wake up. You are needed at the office. In person. It's important." "Now wait a minute, Molly! You've got to be kidding," you grumble. "It's New Year's Day, and I have a hangover. What could possibly be so important anyway?" [Kaku 2011: 353].

This reads like a novel starring you, the reader. As the term "the fictionalized reader" suggests, the technique of fictionalizing the reader contributes to the similarities between popular science and fiction. The chapters or passages similar to the one above could be seen as resembling fanfiction where readers get a chance to apply their imaginations within the story world created by the author. They allow the reader to experience two common, according to Barnes (2015: 73), fanfiction scenarios: "inserting an idealized version of him/herself into the story world, where he or she can interact with favorite fiction characters" and engaging in "'what if' scenarios within the story world."

The fictionalized reader may appear as a curious feature of popular science. However, it serves an important function other than entertaining the actual reader. The technique is an extension of the reader engagement mechanisms, which traditionally do not include presented discourse. With the presence of the fictionalized reader, it is possible to rank presented discourse as a reader engagement strategy alongside reader pronouns, asides, references to shared knowledge, etc. All of these mechanisms are products of the author's skill. So is the fictionalized reader. It is not a reflection of a real reader but an entity entirely made-up by the author.

In the following chapter, I continue to explore the idea of fictionalization when it comes to science and discuss a genre of literary fiction that is firmly rooted in scientific practice.

11

Lab Lit
Fictional Science

Thinking of science as connected to fiction may seem unusual. This is not necessarily a tie that the scientific community embraces wholeheartedly. On the other hand, fiction can be extremely effective at promoting science. So much so that in recent years, that saw a proliferation of scientific and technological advances, literary fiction about science became visible and popular as never before. I am not talking about science fiction but about a different genre called *lab lit*—short for laboratory literature. This genre of fiction has been around for a long time but existed seemingly unnoticed by critics and so overshadowed by science fiction that it went under the radar of many potential readers.

Up until the beginning of the 21st century, there was no name for a genre of literature that dealt with science in its present state, unembellished by the brilliant possibilities of the future. It is a genre that invites a reader to explore the world not in binary terms of either science or literature but in tandem—science as a subject matter and an inspiration *for* literature. The term was coined in 2001 by Jennifer Rohn (2010a: 552), a cell biologist and novelist based at University College London. Lab lit, according to Rohn's (2005) definition, is "a small but growing genre of literary fiction (or other fiction media) in which scientific characters, activities or themes are portrayed in a realistic manner." Lab lit, much like popular science, is about current science or science that is just coming into existence, and when a lab lit author writes about, let's say, an environmental problem—as Barbara Kingsolver does in *Flight Behavior*—her setting is recognizable as the present day. In dealing with today's science, lab lit, similarly to popularizations, often teaches its fundamentals. Bouton (2012), writing for *The New York Times*, notes that in Kingsolver's *Flight Behavior*, the reader "gets some basic ecology."

Another goal of lab lit unites the genre not only with popular science but with professional science as well. In Rohn's (2010a: 552) words, lab lit strives to illuminate and promote "a largely unknown or obscure world" of science. This is why "lab lit tends to focus on the intricacies of scientific work and scientists as people" (Rohn 2010a: 552). The genre and its authors are devoted to making science transparent and showing it as a domain of ordinary humans not an Olympus of the unattainable genius (Pilkington and Pilkington forthcoming).

In trying to present science as a human activity, lab lit is close to popularizations. However, there are many other aspects that tie the two genres. One of them is the focus on new and developing science, the kind of research that the professional mainstream is not yet entirely comfortable with. As Turney (2007: 2) observes, popular science, especially in book form, is uniquely positioned to discuss such emerging concepts as, in physics, for example, "a theory which will unify ... [the micro- and macro-worlds], and account for all the particles and forces which exist, and the properties of space and time, within a single overarching framework." Lab lit does the same thing. Ian McEwan's (1999: 135–136) explorations of the relative nature of time in his novel *The Child in Time* supply a representative example. One of his heroines Thelma, a retired physicist, gives the main character, Stephen, and the reader a rather cynical overview of modern theories of time:

> There's a whole supermarket of theories these days.... One offering has the world dividing every infinitesimal fraction of a second into an infinite number of possible versions, constantly branching and proliferating.... Then there are physicists who find it convenient to describe time as a kind of substance, an efflorescence of undetectable particles. There are dozens of other theories, equally potty. They set out to smooth a few wrinkles in one corner of quantum theory.... But whatever time is, the common-sense, everyday version of it as linear, regular, absolute, marching from left to right, from past through the present to the future, is either nonsense or a tiny fraction of the truth.

McEwan is drawing inspiration for his story line from the untested theories that are occupying the bright scientific minds of today. Tomorrow it might prove all nonsense and appear to future generations of readers to be as ludicrous as, say, galvanism of Mary Shelley's days. But today, there is still the possibility of the truth for any one of those interpretations, and this is what allows McEwan's (1999) work to be regarded as presenting elements of lab lit. Compare McEwan's (1999: 136) conclusion with what Greene (2011: 66), a theoretical physicist and a popular science author, has to say about the nature of time: "Whereas everyday experience convinces us that there is an objective conception of time's passage, relativity shows this to be an artifact of life at slow speeds and weak gravity." Greene (2011: 5–6) also notes that "the familiar notion that any given experiment has one and only one outcome is flawed. The mathematics underlying quantum mechanics—or at least, one perspective on the math—suggests that *all* possible outcomes happen, each inhabiting its own separate universe." There is a direct connection between Greene's (2011: 5–6) statement and Thelma's notion of time that is "constantly branching and proliferating."

Susan Gaines's (2001) novel *Carbon Dreams* is another example of extrapolating from the scientific knowledge available at the time of writing. In the "Author's Note and Selected References," Gaines opens up about the nature of the research her main character conducts: "Tina's actual research projects are fictional, based on sound scientific principles, but with no exact analogues in the real world.... They are ... developed in a credible manner, with results that fit plausibly into gaps in our knowledge and highlight the real trends and upheavals apparent in the scientific literature."

The authors who write lab lit, just like the authors who write popular science, can be professional scientist or those without scientific training. Among the prominent scientific names in lab lit are Jennifer Rohn (a cell biologist), Alan Lightman (a physicist), and Carl Djerassi (a chemist), to list just a few. In fact, writers who choose this genre usually have some kind of connection to the scientific community. For example, Andrea Barrett (whose short story collection *Ship Fever* won U.S. National Book Award) trained as a biologist and attended a PhD program in zoology. Barbara Kingsolver (winner of numerous literary awards) has degrees in ecology and biology and worked as a scientist before her literary fame.

Equally important, however, are the contributions to lab lit by authors without formal scientific training. It might be tempting to wonder if writers like Allegra Goodman, for instance, who did not study or practice science, can present realistic accounts of laboratory life. Can a reader trust them to know what it is like to be a scientist? Yes! Their work is grounded in research on the realities of scientific life in general and on a discipline they are writing about in particular. As David Brin (2016: 18) points out, those authors who plot their stories around scientific developments are quite often "former English majors." These people, "who could not close an equation if their lives depended on it.... Seek out pioneers in any field, plying them with pizza and beer, till they explain something new and wonderful, in terms that any reader could understand." Such authors, in many ways, are closer to their scientifically-inclined lay readers. They know firsthand what aspects of scientific research might pose a challenge and need more background before they will make full sense. As Allegra Goodman (2006) notes in the acknowledgments, she had to learn from scientists "about the care and breeding of mice" in preparation for her novel *Intuition* that deals with a group of scientists working in a cancer research laboratory who come to experience the exaltation of discovery and the consequences of fraud.

If you haven't read lab lit, you might intuitively assume that because this is a genre of literature, its language will be more poetic than scientific and that you will be more likely to learn about relationships among scientists than about details of laboratory research. If you are more interested in science than in scientists, you might feel that you need to reach for a popular science book or article rather than for a lab lit novel or short story. Wait! Lab lit is, actually, much more likely to describe to you the minute technical aspects of laboratory research than any popularization will. In fact, when it comes to some laboratory processes, lab lit is better suited to address the interests of a lay reader than popular science.

The general public tends to associate science with the laboratory. For example, when Rosenthal (1993: 312) conducted a study asking college students who majored in liberal arts what images came to their minds if they thought of scientists, most responded with descriptions of laboratory equipment. Yet traditional popular science media (books, articles, blogs) give very sparse and generic descriptions of the procedures and the apparatus used in everyday laboratory work.

As I note in my article published in *Science as Culture*, "Popular science tends to focus on apparatus that is in some way extraordinary or functions in unexpected ways" (Pilkington 2017: 286). A good example would be the Large Hadron Collider (LHC) of Conseil Européen pour la Recherche Nucléaire (CERN). Here is how Greene (2013) describes it in his article published by the *Smithsonian*:

> Winding its way hundreds of yards under Geneva, Switzerland, crossing the French border and back again, the LHC is a nearly 17-mile-long circular tunnel that serves as a racetrack for smashing together particles of matter.
> The LHC is surrounded by about 9,000 superconducting magnets, and is home to streaming hordes of protons, cycling around the tunnel in both directions, which the magnets accelerate to just shy of the speed of light.

This extraordinary apparatus, however, does not represent the daily realities of laboratory work. And popular science often fails to give the public a glimpse into how ordinary equipment operates. Celebratory presentation of science, which is so common for popularizations (see the Introduction and Conclusion) is an almost exclusively positive portrayal which sometimes results in marginalization of apparatus descriptions in favor of a focus on the grand ideas.

The situation is different for lab lit. In an attempt to be "realistic" and to establish a story world that parallels "actual science culture," lab lit novels have to include aspects "that you and I could easily encounter were we to walk into a research institute, field station or any other place where scientists are doing what they do" (Rohn 2005). As a result, experimental procedures and apparatus associated with them become foregrounded in lab lit (Pilkington 2017: 290–291).

In giving attention to laboratory equipment, lab lit is closer to the everyday realities of professional science than popular accounts are. Consider that human participation in scientific activities is significantly supplemented by the use of apparatus. Some sociologists of science go as far as to propose that no distinction exists between the apparatus and the researcher using it—they are one and the same. As Barad (2003: 829) explains, "'Humans' do not simply assemble different apparatuses for satisfying particular knowledge projects but are themselves specific local parts of the world's ongoing reconfiguring." Bourguet et al. (2002: 7) express the same idea in more familiar terms when they write that scientists desire "to establish themselves as a special entity, combining their own bodies with their experimental tools to make themselves a unique organ for the experience of nature."

There is also the issue of non-human contributors to science that popular science largely ignores and professional science is not too quick to acknowledge either. Lab lit, on the other hand, devotes attention to all the participants in scientific research. Goodman's (2006) *Intuition* is excellent at showcasing the role laboratory mice perform in cancer research and giving the reader an opportunity to learn how a researcher might feel about laboratory animals: "Marion was an attentive and compassionate investigator, almost fond of her small charges, proud and careful of them—not as if they had rights or souls, but as a craftsman might treat precious tools" (Goodman 2006: 23). The interactions between the mice and the scientists are not those between a pet and a caretaker. Throughout the research process, the scientists remain careful and respectful:

> He took just one mouse and put it in the clear plastic container that served as the CO_2 chamber. A simple hose fed into the isolator from a spigot on the wall. Cliff depressed the lever and CO_2 filled the sealed chamber.... Bred for timidity, the little creature still fought death.... The mouse seemed to swell as it expired, growing heavier even as it struggled, until, weighted down, life and color drained. The animal lay still.... Cliff carried the body gently in his gloved hand to the dissecting room. He turned on the examining light and placed the mouse belly-up on the thick polystyrene dissecting block with its disposable pad.... Cliff perched on a stool and went to work.... With tweezers Cliff plucked up the loose pink skin covering the mouse's abdomen, and then

with small sharp scissors he snipped one vertical and four horizontal incisions.... Cliff spilled no blood doing this [Goodman 2006: 68].

In this passage, the attention is on the human participant and his perception of the procedure, but it is also on the mouse and the instruments/equipment. The activities described in this small extract from the novel show not the extraordinary endeavors of a scientific genius and his mighty apparatus but everyday bench work of a typical laboratory researcher. This kind of attention to detail in an unremarkable process is unlikely in popular science but usual in lab lit. It supplies detailed descriptions of equipment no matter its size, appearance, or importance.

Routinely, in "these novels apparatus is used as a means of characterization for the scientists, and it is not unusual to have apparatus itself treated as a character" (Pilkington 2017: 286, 298). Gaines (2001), for example, uses laboratory equipment to describe the emotional state of her heroine:

> ... when she [Tina] opened the door of the instrument lab and heard the pump, she realized that the HPLC had other plans, and her heart sank.... Sometimes she felt like the thing was alive, manipulating her every move.... She almost wished it would explode. But, of course, it would never do anything that heroic. It would just burn out, leading her on, tormenting her to the very end [36].

At first, it appears that the author is focusing on the instrument; however, a careful reader will notice that "the scientist endows her equipment with human emotions by projecting her own. At the point when the apparatus malfunctions, it is Tina's emotions that are presented in terms of the apparatus." She might want the machine to "explode," but it is she who is ready to burst in anger and frustration. The equipment might seem "alive" to her, but she is the only living creature in the laboratory, and while she accuses her apparatus of being manipulative, "she is the one who physically manipulates the machinery. In this instance, the apparatus is both a character and a foil for the main" character (Pilkington 2017: 301).

Jennifer Rohn (2010b: 58, 64, 65) uses a similar technique in *The Honest Look*, but she goes one step further and assigns unmistakable human qualities to the equipment her researcher uses: "She fed the [machine] a few sips of buffered saline, then twiddled a knob to burp the air out of its labyrinth of flexible latex microtubing." This apparatus proceeds to act like a human baby. It displays content when the "feeding" continues: the machine "emitted a self-satisfied bleep and a sprinkling of flashes on the console." And it cries when something is not to its liking: the apparatus "began to ping in panic." This is no longer a projection of the scientist's feelings—the heroine is not showcasing her maternal instinct—instead this is a sustained effort to treat laboratory equipment as equal to the scientist who works with it—both are human. In this regard, lab lit is ahead of popular science in recognizing the important role of the non-human contributors to scientific progress.

As we discuss the advantages of lab lit, it is important to keep in mind that, as a branch of science writing, the genre shares the celebratory goal with professional and popular science. One of the ideas behind lab lit is to show scientists as engaging and interesting people. Lab lit authors create characters with whom multiple readers can find similarities. These are not books for scientists; this is a literature for the lay public. Lab lit repeatedly stresses the point that science and scientists function not inside an enclosed group but as part of society. Lab lit draws connections between the scientific culture and the diverse cultures that makeup the human family. Lab lit's characters, just like real scientists, write and read poetry, appreciate art, and listen to rock and classical music while performing experiments. In *The Honest Look* (2010b), for example, Rohn plays out the tension between science and literature through the personal struggles of her main character, Claire Cyrus. It is only when she lets herself be both a scientist and a poet that Claire finds her equilibrium. A.S. Byatt (2005: 294) also makes a point of recognizing the artistic and the logical as continuations of each other. She writes of her characters in *The Virgin in the Garden*, "My heroine, brooding about seventeenth-century metaphors ... and my mathematician, were in fact struggling with the same problem" (294).

An encounter with lab lit is, for many readers, a point when they realize that scientists are people—they have families and friends; they raise children and take care of elderly parents; they face the problems of the everyday with the same vigor or resentment that a person in any other profession would. Lab lit is written not only to promote a positive image of science and its practitioners but also to tell interesting stories about people who are connected to science. Lab lit is first and foremost about people: it presents scientists not as mysterious keepers of Nature's secrets but as human beings with whom a reader can easily relate. Lightman (1993) in *Einstein's Dreams* stays true to presenting the complexities of the theory of relativity, but he also shows his readers the great scientist as a friend and family man. Was Einstein an ideal husband, a loyal friend? The reader gets to draw her own conclusions:

> Einstein and Besso sit at an outdoor café.... It is noon, and Besso has talked his friend into leaving the office and getting some air.... "I'm making progress," says Einstein. "I can tell," says Besso, studying with alarm the dark circles under his friend's eyes.... Besso remembers when he looked just like Einstein does now, but for a different reason. It was in Zürich. Besso's father died suddenly.... Besso, who had never gotten along with his father, felt grief-stricken and guilty. His studies came to a halt. To

Besso's surprise, Einstein brought him into his lodgings and took care of him for a month [Lightman 1993: 97–98].

Here is one of the instances where Einstein's marriage is discussed in the novel:

> He [Besso] is puzzled why his friend ever got married in the first place. Einstein himself can't explain it. He once admitted to Besso that he had hoped Mileva would at least do housework, but it hasn't worked out that way. The unmade bed, the dirty laundry, the piles of dishes are just as before. And there have been even more chores with the baby [Lightman 1993: 100].

Throughout the novel, Einstein is often referred to as a friend, and the whole narrative is formed around his relationship with Besso. This is not an accidental choice. In many ways, lab lit attempts to undo the effects of the alienation between the public and the scientific community. Stressing the relationships outside the lab is one of the ways in which lab lit connects the scientific community with the world of the readers.

At the same time, this is a genre that invites the non-professionals inside the laboratory and shares the secrets and hopes of the scientific community. Lab lit brings to the forefront not only contemporary scientific ideas (as popular science does) but also the emotional issues of competitiveness among scientists, the difficulties of getting funding, the dedication to one's research goals, and the ethical dilemmas of using human or animal subjects, to name just a few. These works allow the reader to co-experience what it is like to be a scientist.

Reading a lab lit novel, a play, or a collection of poems will not make the readers scientists, but for the duration of the work, they will be immersed in the world of the scientific community and will have to ponder the same questions and make decisions about the same problems as the scientists who inhabit the fictional laboratory. Inevitably, this journey to the laboratory will create a deeper understanding and a sense of connection with the real world of science. For example, when the reader is exposed to detailed descriptions of laboratory equipment and given detailed representations of what experimental procedures look like, he has an opportunity to form an opinion about the work that scientists do. A reader knowledgeable in the basics of research design and experimentation will be less susceptible to anti-science rhetoric and will, possibly, become an ally of the scientific community. In offering this kind of access to the research process, lab lit stands apart from popular science. In a way, this genre of literature is the only available option for a lay person to get a look inside the laboratory. It is a unique branch of science writing that deserves appreciation and recognition.

Conclusion: Professional Science and Popular Science

In chapter 1 and in the introduction, I mentioned that the notion of professional science, as opposed to popular science, is a relatively recent construct. While the scientific method has been around since the times of the Ancient Greeks, the idea that only a select few can practice it dates back to 19th-century Britain. Before then, the sciences were potentially accessible to any interested person and, in fact, attracted many amateurs who ended up making scientific contributions. The welcoming of amateur scientists was, according to Coppola (2016: 3), a calculated economic and political move on the part of Charles II, who "saw the potential for real material benefits, or in the very least an inoffensive diversion, in the tireless and omnivorously curious pursuits of the so-called virtuosi, the non-professional gentlemen-scientists who filled the ranks of the Royal Society and read the *Philosophical Transactions*, amateurs who experimented for no reason other than a virtuous love of truth, and to beguile their leisure hours in ingenious pursuits."

Popular science was born not out of great necessity but as a result of a power struggle. Ironically, the deliberate actions that separated professional science from the popular variety tied them together more tightly than any organic parting of the ways would have accomplished. In an attempt to control both (the professional and the popular outlets) the scientific community forever solidified its connection with the world of amateurs. For a long while, popular science authors remained second-class citizens compared to the exalted inhabitants of the professional laboratories. Today, however, the crossover from a professional scientist to a popular science author is as desirable as a sturdy lock on the doors of the 19th-century lab. Think of Neil de Grasse Tyson, Michio Kaku, Lee Smolin, or Brian Greene. Their popularizing careers certainly did not bring them any shame, nor did their popularity hurt the scientific research they conduct.

If anything, having access to an interested audience and to additional dissemination channels proved to be tremendously beneficial. One outcome of popularization is name recognition, and not only for the scientists but also for their research projects. The notion of popular science as separate or different from "real" science is outdated. Professional and popular science share an audience, linguistic elements, structural organization of their texts, and more importantly, they share a common goal—to promote science and to present it in the best possible light.

In chapter 4, I demonstrated specific linguistic techniques that authors of popular science can use to create and distribute a benevolent image of science. Mainly, I suggested that using certain narrative patterns automatically predisposes a popular science text to a positive outlook on the content. A certain degree of subjectivity is expected of popular science; after all, these books, articles, and blog posts embrace and celebrate human participants in research. As Parkinson and Adendorff (2004: 388) point out, "popular articles focus on people," while research articles pay more attention to methodology. At the same time, the relationship between professional and popular science is not that clear cut. Their shared roots never allow them to get too far away from each other. In certain instances popular science emphasizes research methods and apparatus, and quite often professional publications turn to narrative techniques that place emphasis on researchers and their extraordinary accomplishments. More importantly, there is a shared language.

Once practicing science professionally became a privilege, those with the access to the laboratory set out to create and promote a specialized, jargon-laden language designed to code the scientific findings sufficiently so that only the members of the inner circle would be able to make sense of them. Topham (2000: 561) calls the emergent language of professional science "increasingly arcane and technical."

This linguistic legacy still impacts both professional and popular science. The particular phrasing and sentence structures of modern research articles are designed to support the 19th-century notions about protecting scientific information from laymen. Extensive use of the passive voice, nominalizations, and discipline-specific terminology create an effect of objectivity for a professional scientific text. Let me show you why the objectivity is illusory.

First of all, the very structure of a research article is such that it does not allow for negative outcomes. A research article is essentially a narrative even though the use of narrative technique was supposedly expunged during the professionalization of the scientific community. Olson (2015: 8) points out that "science is a profession that is permeated with narrative structure and process, yet scientists are so blind to the importance of narrative." He writes off this blindness as "'storyphobia'" without examining its roots, but he does suggest that "science is a newly arrived guest in an ancient narrative world" (Olson 2015: 8, 37). In fact, popular science still draws heavily on narrative's powers of explanation and persuasion (see chapters 3 and 4)—a strategy abandoned by professional scientists. We know from the history of science that the use of narrative was marginalized through deliberate actions and exclusion of those practitioners who were likely to use narrative presentation of their research. Refusal to use narrative techniques openly was one more step in the direction of professionalization. Modern scientists inherited the negative attitude to storytelling and underwent training that specifically told them to avoid narrative technique in their writing. So it is not surprising that they fail to see the similarity between a traditional narrative and a research article.

Recognizing the narrative nature of science presentation would effectively undermine the scientific community's exclusive access to truth. That is why linguistic studies that reveal the structures and explain the language of professional and popular publications are so important. As Calsamiglia (2003: 141) observes, linguistic studies of how science is communicated in written and oral forms have the ability "to remove many of the preconceptions—not to say prejudices—that people have about science." She continues, "Such ill-founded notions include the fact of considering science to be a neutral activity; the fact that science procures stable and eternally valid truths; the fact that the scientific community is licensed to account for natural, human and social reality; the fact of the 'hallowing' of scientific knowledge and its 'priests'; the fact of conceiving the linguistic representation of science as rhetoric free, maximally informative and transparent; the fact that science is a pure activity, out of touch with market forces and politics, etc." (Calsamiglia 2003:142).

Another, and perhaps less obvious, point of connection between professional and popular science has to do with imagination—or fictionality, as those who study narratology refer to it—or modeling, the term used by the philosophers of science. Generally, it includes the use of projections, predictions, imaginary scenarios, and emotionality. The discussion of fictionality in connection with professional science is usually not a welcome topic in professional circles because it somewhat undermines the message that science deals exclusively in ultimate truths and solid facts.

During the 18th and the 19th centuries, at the time of the onset and the development of professionalization, dissemination of scientific results to colleagues and promotion of scientific achievements to the public often involved theatricality and visual rhetoric, which today is firmly associated with literary fictions, not with professional science. Yet, as Myers (1997: 51) observes, even the Bakerian Lectures delivered at the Royal Society and "published in the *Philosophical Transactions*, reviewed in the quarterlies" relied on rhetoric of theatricality. Of Humphry Davy's 1808 lecture Myers (1997: 49) writes that Davy set up "a situation for telling based on the rhetoric of showing." This is accomplished by the "attribution of agency to humans or equipment, his use of adjectives…, and his anticipation of objections." In that lecture Davy combines the familiar structure of an experimental account, complete with the use of the passive voice, and a demonstration—essentially, a popularizing technique. "It is more like some popularizations in its concern with visual effect," Myers (1997: 51) concludes. At the same time, the audience for Davy's lecture was not the lay public but "the elite of the Royal Society … expected to be able to draw on a shared background in chemistry, to follow his reasoning and do without his spectacular shows" (Myers 1997: 51).

Coppola (2016: 9) confirms the connection between science and visual presentation of new knowledge and new technologies. "Science was turned to spectacle," he notes, and "unambiguously offered as entertainment," which was conceived, produced, and performed by scientists themselves. Henry Pepper's ghost lecture at the Royal Polytechnic Institute is a representative example: "Henry Pepper conjured up on the stage by virtue of an elaborate hidden optical apparatus" a projection, which he "would exhibit … with pomp and mystery, in a lecture that subsequently divulged the secrets of the contrivance to the audience. Then … Pepper would conclude the evening's entertainment by featuring his ghost as a show-stopping special effect in the performance of a gothic play or popular opera" (Coppola 2016: 8). Such spectacular means of science presentation, especially to professional audiences, needed to be overcome in order to preserve science as the elevated activity of a select few. Print media, where the experimental findings could be sufficiently coded, provided a more welcome outlet for professional dissemination. Fictionality, with its spectacular theatrical effects, had no place in professional science, and its presence was significantly limited in popularizations as well.

Conclusion

First introduced by Hans Vaihinger, a German philosopher, in 1924, the idea of fictionality extending beyond literature—into non-fiction and especially into science—was shelved until 1993. At that time, Arthur Fine revived the interest in the concept, which inspired further research (see, for example, Suarez 2009, Toon 2012, Barwich 2013). Today, fictionality, from the point of view of professional science, is usually understood as "the role played by particular methods of model building such as abstractions, idealisations, and the employment of highly hypothetical entities" (Barwich 2013: 357–358). This view of fictionality focuses on an important aspect of the scientific process—hypothesizing. Philosophers of science (see, for example, Cartwright 1983, Toon 2012) have circulated the idea that fictionality in science is "analogous to literary ... fiction" (Rouse 2009: 37). Toon (2012) devotes a book-length study to demonstrating a connection between solutions to scientific problems and what he calls "make-believe"—the idea of using theoretical and physical model scenarios. Titled *Models as Make-Believe: Imagination, Fiction and Scientific Representation*, it fully embraces the presence of fictionality in professional science and demonstrates its usefulness. Similarly to the situation with narrative, the denial of a connection between science and fiction by the scientific community does not automatically equal the absence of the link.

"Why would scientists refuse to acknowledge certain aspects of their disciplines?" you might ask. Think about how science got to be an area of inquiry separate from other forms of knowledge. Professionalization involved not only the exile of amateurs but also the severing of connections with other disciplines that did not rely on the scientific method to obtain their knowledge about the world. Narrative technique and fictionality could not possibly be part of the scientific process. Recognizing their contributions would have undermined the status of professional science as the only form of objective knowledge. However, the human brain is predisposed to learning from stories and imagination. The denial of these components in professional science might have been effective, but it also proved detrimental.

Modern scientists are desperately trying to reestablish the connection with the public as funding and prestige depend as much upon social recognition as they do upon acknowledgment of research by colleagues. Using the coded language of professional science used to assure that no amateur would gain access to the laboratory; today it ensures that grant-providing agencies have no clear idea what their money will be used for and will choose to support more familiar endeavors or even relegate their funds to anti-science movements that easily distort the scientific messages in the process of converting the language of the scientific community into human-speak.

The popularity of communication books for scientists is not accidental. Some of them go as far as to suggest that in order to talk or write to non-specialist audiences scientists have to undergo drastic changes in their attitudes toward the lay public (see, for example, Barton 2010: 103–123, Bowater and Yoman 2013: 85–90). Science is reaping the fruits of its isolationist policy. Meredith (2010: 18) and Sackler (2014: 8–10) demonstrate a concern that the public does not always see the members of the scientific community as "warm" (Sackler 2014: 10). The lay people want a personal connection, which the scientific community fought so adamantly to expunge. Sackler (2014: 10) proposes that scientists would be better received by the public if they focus on such personal and emotional aspects "as letting people know why someone went into science."

While the professional scientific community struggles to find its way to the hearts and minds of non-specialists, popular science remains their best ally. This outlet of the scientific community never abandoned narrative and fictionalizing techniques, and today it is proving more effective than ever. While professional science may never fully embrace the idea of using narrative and fictionality openly, nowhere is the connection between fiction and science more obvious than in popularizations.

In chapters 5 and 6, I addressed the roles the speech and thoughts of scientists play in dramatizing popular accounts of discoveries. In fact, a more detailed look at the presented discourse of professional scientists in popular books suggests that all instances of the scientists' speech or thought contain a degree of dramatization and thus contribute to fictionality (Pilkington 2018).

The most famous example of fictionality in professional science is probably the Schrödinger's cat thought experiment—an entirely imaginary scenario that, it seems, acquired a life of its own, becoming part of popular culture as both an opportunity to reference serious scientific research and an invitation to science-themed jokes. Note that the thought experiment was suggested by Schrödinger (1935) in a letter to Einstein (as a discussion of an article Einstein co-authored) not as part of a professional publication. Modeling, both theoretical and physical, is an intermediate step between a research question and a publishable answer; as such it usually is left out of professional publications that focus on the results not on the processes of obtaining them. Popular science, on the other hand, embraces these imaginary scenarios. I discuss thought experiments in popular science in chapter 10 and show that they are some of the most interactive segments in popularizations, using stories and games. These are also portions of the texts that present the reader as a character in imaginary situations.

Fictionality and the use of narratives are not the only points of connection between professional and popular science. Perhaps the most obvious similarity is the use of scientific terminology. Here too, professional and popular science have shared history.

Scientific terminology is not limited to professional research articles. I devoted chapters 8 and 9 to the discussion of definitions of scientific terminology in popular science. In fact, for popular science, scientific terms presented a problem and an opportunity from the very beginning. As codifying scientific results through extensive use of esoteric terms became a common and desirable practice, the 19th century saw a rise in popular science texts in a form of dialogue. Why dialogue and not narrative? Because, as Myers (1992: 238) explains, printed dialogues that popularize scientific findings bring to the forefront the issue of language in a way that narrative structure does not allow. Dialogues predispose the participant to asking questions—whereas narratives have to go out of their way and employ special techniques to allow for such interactions to be successful (see chapter 10). Dialogues, according to Myers (1992: 238), focus on explanations of terminology and make sure the reader understands the basic terms before moving on. By doing that, popular science in dialogic form acknowledges the difficulties the language of professional science poses for lay readers. Other forms of popular texts do that as well, but, in the words of Myers (1992: 238), dialogues "teach us to read conventional scientific texts by making explicit what is usually implicit.... Stepping back from the text involved in dialogue allows for discussion of terms."

These "discussions of terms" were increasingly necessary as the scientific community was deliberately setting itself up as a separate linguistic, as well as intellectual, entity. Those popularizers who were outside the elite circle of professional scientists had to learn and practice the new language along with their readers. Today, the issue of scientific terminology in popular science is still vital as numerous popularizers try to come up with creative and effective ways of incorporating the jargon of the various scientific disciplines into discourse designed primarily for non-specialists (see chapters 8 and 9).

As we have seen, popular and professional science are knowledge-disseminating outlets of the scientific community. Their shared history and the many linguistic attributes that unite them demonstrate not necessarily the superiority of professional science as a form of human inquiry but rather its rightful place among the *various* ways in which we explore our environment. Science, be it in professional or popular form, is not an unprecedented mechanism for acquiring knowledge; it is a logical result of the centuries of human development, and it caries with it our predispositions (narrative), our imagination, and our hopes (fictionality). It is, at the very least, unproductive to deny those aspects of science that merge it with the common human experience.

Epilogue

In this book, I talk about the broad language of popular science, yet all of the examples and most of the studies I cite deal with the English language. How could an investigation of the linguistic properties of popular science in English alone afford such a generalization? For the most part, it can't, and the findings discussed in the book apply, first of all, to English-language popular scientific writing. However, English is so powerful as a language that general conclusions are not completely out of the question: "There are clear indicators in the present of English's dominance over other languages. Former colonies of the British Empire, countries and areas currently under U.S. control, and even the former republics of the Soviet Union all are establishing their own, distinct linguistic relationships with the English language. It has been common among linguists to talk about 'Englishes' as an acknowledgment of the contribution that non-native speaking countries make to the English language" (Pilkington forthcoming).

In the last few centuries, English has also emerged as the language of science. At the same time, as I point out in my book *Presented Discourse in Popular Science*, "Historically ... English has not always been the language of science even in the English-speaking world, let alone internationally" (2018: 5). The scientific community, being an international group almost from the beginning, always looked for a single language that all of its members would share (and, no, that language was not mathematics). In their linguistic preferences, scientists are pretty much anti-diversity. As Gordin (2015: 2) puts it, scientists are "the most resolutely monoglot international community the world has ever seen." The first language of science (at least in Europe) was Latin. It was the language of scientific choice from the middle ages up until the mid–17th century. Later on, several Indo-European languages (Russian, French, English, German) and Japanese each dominated scientific publications from about 1880 to 2005 (Gordin 2015: 6).

If we consider the language of science in its professional and popular manifestations, we are inevitably talking about a written language. Scientific results are disseminated not by word of mouth (though conferences play an important role) but through peer-reviewed publications. In fact, for many historians and sociologists of science, the research article is equated with scientific results. For example, "Latour and Woolgar (1979) say that articles are the final products of a laboratory … [and] acknowledge that ideas and knowledge (embodied in texts) are the results of the scientific process" (Pilkington 2017: 288). Souba (2011: 57) goes so far as to propose that a text can be equal to a phenomenon: "A scientific paper whose results are largely accepted by the relevant scientific community is a discovery." With that in mind, it is curious that popular science authors choose to emphasize the speech and thoughts of scientists and do not include as many references to writing. They may prefer presentation of speech and thought, in part, because this preference allows for further separation from professional scientific discourse, which some readers might consider confusing and not engaging.

Throughout the history of the language of science, this separation was not always limited to certain linguistic features. There was a time when the language of the professional scientific community and the language of popular science were literally different languages. Early on, English played a significant role as a language of popularization (at least in the English-speaking world). Even at the time when Latin was the go-to language of professional science, there was already an understanding that non-professionals needed a modified version of scientific discourse. One instance of such modification is illustrated very well by none other than Geoffrey Chaucer.

Circa 1391, Chaucer wrote *A Treatise on the Astrolabe*. In the words of Seth Lerer (2015: 80) it was "a synthesis of medieval astronomical and astrological teaching inherited from Greek and Latin, Arabic and European teachings." However, Chaucer had the idea of simplifying "a technical language" of his Latin original. In essence, he set out to compose a popularization. The modification that he introduced was not just in retelling the Latin text in a more relatable manner. He used an entirely different language for his work. The audience for the *Treatise* was Chaucer's ten-year-old son, Lewis, to whom Chaucer hoped to "reveal … in … easy English the conclusions concerning this material in as true a fashion as any ordinary treatise shows in Latin" (transl. from Middle English by Lerer 2015: 81). Chaucer does this because Lewis is too young to understand Latin well enough, yet he is obviously old enough to be exposed to the science of the

Treatise. To my knowledge, this is one of the earliest practical realizations of scientific knowledge in a language not common among scientists.

In the introduction to his text, Chaucer makes sure to note that he is using English "only for [his son's] instruction" (Lerer 2015: 82). He does that in order not to offend the scientific community and to appear to preserve their right to a linguistic code that only a few can understand. At the same time, Chaucer recognizes a basic goal of a popularization (to make science accessible) and the inaccessibility of the language of science to those who are not specially trained in its usage and sets out to make a change. This is one of the first steps of the English language on the path to being the universal language of science, and it started with being the language of popularization.

Today, as the unquestionable favorite among professional scientists, English also remains influential in the realm of popularizations. The influence of the English language goes beyond content; that is, popular science written in other languages uses linguistic features of English-language popularizations even when the texts are original and not translations from English. In general, it is not surprising that English is influencing other languages. As Kranich et al. (2012: 315–316) note, most of the texts translated today are translated "from English, the dominant *lingua franca*." American popular culture, the rise in technological advancements, and the increased globalization of the world all contribute to a certain shared vocabulary and result in some degree of familiarity with the English language even by those who do not study it purposefully (Pilkington 2011). Kranich et al. (2012: 316) point out that "the dominance of the English language in today's global communication leads to variation and change of indigenous communicative norms." For popular scientific writing, the "variation and change" mean increased linguistic homogeneity. In other words, popular science in other language begins to sound like English.

Kranich (2016: 165) asks if "translations [can be seen] as trigger of linguistic change." Her evidence points to a positive response. For example, Kranich et al. (2012) and Kranich (2016) note that interactivity of English-language popular science is a feature that gets mimicked by other languages less prone to interactive exchanges in writing. Popular science written in German and translated from English into German became more interactive with time. Kranich (2016: 165) lists studies that document the process and "present evidence for an increasing interactionality of English-German translations as well as of German non-translated popular science texts." According to Kranich et al. (2012) and Kranich (2016), some of the features common in English-language popular science that became used more frequently in German include starting sentences with "And" (*Und*), using the pronoun "we" (German *wir*, which is considered a reader-inclusive pronoun in popular science and is usually used to represent the author and the reader), the use of "but" (German *aber*, which allows the author to introduce an alternative point of view). That is not to say that German did not employ these constructs before English popularizations came along; English-language popular science simply promoted certain ways of organizing information, and over time, even those writers who composed their popularizations in German (as opposed to translating from English) began to use these methods more and more frequently.

When certain elements of the original language are preserved in a translation (even if they are uncommon in the language into which a translation is made) it is called "shining through." Over time, as Kranich's research clearly shows, what begins as a shining through can become a permanent feature of the target language. In this way, a look at linguistic features of English-language popular science is an investigation of potential universals. Of course, not all peculiarities of popular scientific texts written in English will one day become the norms of other languages, but the elements that have to do with the interactions between the reader and the writers are very likely candidates. That means that the fictionalized reader or definitional strings, for example, might become key features of popular science texts written in Russian of French.

As Kranich el al. (2012), Kranich (2016) and Mair (2006) show, interactivity and general tendency for informality are characteristic features of English-language popular science. These features are a result of cultural trends towards "the democratization of knowledge" (Kranich et al. 2012: 332). Proliferation of popular science in general fits this model of a more interactive and inclusive approach to information gathering and processing.

As I say in the introduction, the idea that science, and especially scientific discovery, is the domain of professionals has been accepted by our society. For instance, Bensaude-Vincent (2001: 100) writes that "when it is assumed that the advancement of science is a natural and necessary process that nothing—no human intervention—can stop, then nothing can prevent the increasing gulf between the professional scientists in charge of the production of knowledge and the public that consumes the products of knowledge." However, technological advancements and, as a result, complicated apparatus that were supposed to keep the curious general public out of the laboratory have had the exact opposite effect.

Democratization of knowledge also means democratization of technology and research methods. Modern scientists, quite often, invite the public to contribute data that may at some point produce a discovery. The roles of the scientist as a producer of science and the public as a consumer shifted once technologies like CRISPR, to introduce one example, entered the scene. Here is what one of its developers, Dr. Jennifer Doudna (2017), speaking at the Aspen Ideas Festival, had to say about it: "One of the things that's so on one hand wonderful but also very challenging about this technology is that it's widely available, right, whether patent offices notwithstanding, anybody doing academic or even commercial research right now can easily get a hold of the CRISPR molecules and tools for doing gene editing and they can do it. And that's of course happening." However, there appears to be no requirement that a purchaser of the gene-editing kit be a scientist, and there certainly does not appear to be any vetting process for those who wish to purchase the product. One supplier of CRISPR kits is OriGene, a website where for around $1300 anyone can purchase a "human gene knockout kit via CRISPR." Nowhere in the Legal Notices does it say that a potential buyer has to be a professional scientist. Anyone with access to a lab can try out this technology and potentially contribute to scientific progress.

Another example of the public's productive involvement in scientific discoveries comes from the field of mathematics. With the help of the Great Internet Mersenne Prime Search, or GIMPS, anyone can discover the next largest prime number. "The idea of this software is to utilize a computer's idle time to do computations" (du Sautoy 2011: 46). The most recent discovery of a Mersenne prime was done using this software, and the discoverer was not a professional scientist but a deacon in the Memphis suburb of Germantown, who used a church computer with GIMPS software (Prashadjan 2018).

What do these developments mean for the language of professional and popular science? As some linguists have already pointed out, the language of science and especially the language of popularizations is becoming more interactive, more aware of the reader, who could end up making contributions to the next big discovery. It also means that those concerned with the language of science in all its manifestations pay more attention to and uncover more features that contribute to interaction. Already such studies exist. Livnat (2012) writes about dialogicity of professional scientific writing. "The scientific text or what is often called academic prose is often presented on the monologic-dialogic continuum as a classic example of a monologic text…. However, a more in-depth scrutiny of the nature of scientific writing will show that it is only partially monologic in character"

(Livnat 2012: 1). Her study concludes that "a scientific project is constructed step by step by means of a dialogue with its readers and discourse community" (Livnat 2012: 2).

This does not, however, mean that the formal structure and jargon of the research article will be abandoned. There are slight modifications, and there is a call to use active voice more frequently, for example, in order to make scientific writing more similar to conventional use of the English language (Millar et al. 2013). In other words, the language of science is very, very slowly moving closer to the language of popularizations. And the global language of popular science is moving closer to the norms and preferences of the English language. The conclusions of this book, therefore, while drawn from the English-language examples exclusively, could potentially be relevant to explorations of the language of popular science in general.

Sources

List of Primary Texts
Bryson, B. (2003). *A Short History of Nearly Everything.* New York: Broadway Books.
Bynum, W. (2012). *A Little History of Science.* New Haven: Yale University Press.
Carroll, S. (2012). *The Particle at the End of the Universe: How the Hunt for the Higgs Boson Leads Us to the Edge of a New World.* New York: Plume.
Coen, E. (2012). *Cells to Civilization: The Principles of Change That Shape Life.* Princeton: Princeton University Press.
du Sautoy, M. (2010). *The Number Mysteries: A Mathematical Odyssey through Everyday Life.* New York: Palgrave Macmillan.
Ferris, T. (1988). *Coming of Age in the Milky Way.* New York: William Morrow.
Gaines, S.M. (2001). *Carbon Dreams.* Berkeley: Creative Arts.
Goodman, A. (2006). *Intuition.* New York: The Dial Press.
Greene, B. (2011). *The Hidden Reality: Parallel Universes and the Deep Laws of the Cosmos.* New York: Alfred A. Knopf.
Kaku, M. (2011). *Physics of the Future: How Science Will Shape Human Destiny and Our Daily Lives by the Year 2100.* New York: Doubleday.
Kean, S. (2012). *The Violinist's Thumb and Other Lost Tales of Love, War, and Genius, as Written by Our Genetic Code.* New York: Little, Brown.
Lightman, A. (1993). *Einstein's Dreams.* New York: Pantheon.
McEwan, I. (1999). *The Child in Time.* New York: Anchor.
Rohn, J. L. (2010b). *The Honest Look.* New York: Cold Spring Harbor Laboratory Press.
Zimmer, C. (2011). *A Planet of Viruses.* Chicago: University of Chicago Press.

References
Ahbel-Rappe, S. (2009) *Socrates: A Guide for the Perplexed.* London: Continuum.
Ainley, M., M. Corrigan, and N. Richardson. (2005). "Students, tasks and emotions: Identifying the contribution of emotions to students' reading of popular culture and popular science texts." *Learning and Instruction* 15: 433–447.
Aristotle. (1984a). "Physics" in Jonathan Barnes (ed.), R.P. Hardie and R.K. Gaye (transl.), *The Complete Works of Aristotle: The Revised Oxford Translation Volume One,* 315–446. Princeton: Princeton University Press.
Aristotle. (1984b). "Posterior analytics" in Jonathan Barnes (ed.), Jonathan Barnes (transl.), *The Complete Works of Aristotle: The Revised Oxford Translation Volume One,* 114–166. Princeton: Princeton University Press.
Avraamidou, L., and J. Osborne. (2008). "Science as narrative: The story of the discovery

of penicillin." *The Pantaneto Forum* 31. Accessible: http://www.pantaneto.co.uk/issue31/avraamidou.htm Accessed: 11 May 2015.

Baerger, D. R., and D. P. McAdam. (1999). "Life story coherence and its relation to psychological well-being." *Narrative Inquiry* 9/1: 69–96.

Balfe, M. (2007). "Diets and discipline: The narratives of practice of university students with type I diabetes." *Sociology of Health and Illness* 29: 136–153.

Barad, K. (2003). "Posthuman performativity: Toward an understanding of how matter comes to matter." *Signs* 28/3: 801–831.

Barnbrook, G. (2002). *Defining Language: A Local Grammar of Definition Sentences*. Philadelphia: John Benjamins.

Barnes, J. L. (2015). "Fanfiction as imaginary play: What fan-written stories can tell us about the cognitive science of fiction." *Poetics* 48: 69–82.

Barton, N. (2010). *Escape from the Ivory Tower: A Guide to Making Your Science Matter*. London: Island Press.

Barwich, A.-S. (2013). "Science and fiction: Analysing the concept in science and its Limits." *Journal for General Philosophy of Science* 44: 357–373.

Baumgarten, N., and J. Probst. (2004). "The interaction of *spokeness* and *writtenness* in audience design." In House, J., and J. Rehbein (eds.), *Multilingual Communication*, 63–86. Philadelphia: John Benjamins.

Bell, A. (1984). "Language style as audience design." *Language in Society* 13/2: 145–204.

Bell, A. (1991). *The Language of News Media*. Oxford: Blackwell.

Bensaude-Vincent, B. (2001). "A genealogy of the increasing gap between science and the public." *Public Understanding of Science* 10: 99–113.

Bergenholtz, H., and S. Tarp (eds.). (1995). *Manual of Specialized Lexicography: The Preparation of Specialized Dictionaries*. Philadelphia: John Benjamins.

Berman, R. A. (1997). "Narrative theory and narrative development: The Labovian impact." *Journal of Narrative and Life History* 7/1–4: 235–244.

Blanchard, L., V. Bhaskar, and L. Briefer. (2015). "The lost narrative: Ecosystem service narratives and the missing Wasatch watershed conservation story." *Ecosystem Services* 16: 105–111.

Bourguet, M., C. Licoppe, and O. Sibum. (2002). *Instruments, Travel, and Science: Itineraries of Precision from the Seventeenth to the Twentieth Century*. London: Routledge.

Bouton, K. (2012). "In lab lit, fiction meets science of the real world." *The New York Times*. 3 Dec. Accessible: http://www.nytimes.com/2012/12/04/science/in-lab-lit-fiction-meets-science-of-the-real-world.html?_r=0 Accessed: 17 April 2018.

Bowater, L., and K. Yeoman. (2013). *Science Communication: A Practical Guide for Scientists*. Chichester: Wiley-Blackwell.

Boyd, B. (2008) "The art of literature and the science of literature." *The American Scholar* Spring. Accessible: http://theamericanscholar.org/the-art-of-literature-and-the-science-of-literature/#.UrMm7uLZgYI Accessed: 20 February 2016.

Boyd, B. (2009). *On the Origin of Stories: Evolution, Cognition, and Fiction*. Cambridge: Belknap Press.

Brin, D. (2016) *Insistence of Vision*. Lanham, Maryland: The Story Plant.

Bucchi, M. (1998) *Science and the Media: Alternative Routes in Scientific Communication*. London: Routledge.

Byatt, A.S. (2005). "Fiction informed by science." *Nature* 434: 294–296.

Calsamiglia, H. (2003). "Popularization discourse." *Discourse Studies* 5/2: 139–146.

Calsamiglia, H., and C.L. Ferrero. (2003). "Role and position of scientific voices: Reported speech in the media." *Discourse Studies* 5/2: 147–173.

Calsamiglia, H., and T. Van Dijk. (2004). "Popularization discourse and knowledge about the genome." *Discourse & Society* 15/4: 369–389.

Cartwright, N. (1983). *How the Laws of Physics Lie*. Oxford: Oxford University Press.
Chakrabarti, K. K. (1995). *Definition and Induction: A Historical and Comparative Study*. Honolulu: University of Hawaii Press.
Chivers, T. (2010). "Popular science books take off: A big bang in physics publishing." *The Telegraph*. 6 Sept. Accessible: https://www.telegraph.co.uk/culture/books/7985508/Popular-science-books-take-off-a-big-bang-in-physics-publishing.html Accessed: 26 May 2018.
Ciapuscio, G. (2003). "Formulation and reformulation procedures in verbal interactions between experts and (semi-)laypersons." *Discourse Studies* 5/2: 207–233.
Cohn, D. (1990). "Signposts of fictionality: A narratological perspective." *Poetics Today* 11/4: 775–804.
Copi, I. M. (1972). *Introduction to Logic*. New York: Macmillan.
Coppola, A. (2016). *The Theater of Experiment: Staging Natural Philosophy in Eighteenth-Century Britain*. Oxford: Oxford University Press.
Culler, J. (1982/2008). *On Deconstruction: Theory and Criticism after Structuralism*. Ithaca: Cornell University Press.
Curtis, R. (1994). "Narrative form and normative force: Baconian story-telling in popular science." *Social Studies of Science* 24: 419–461.
Dahl, T. (2015) "Contested science in the media: Linguistic traces of news' writers framing activity." *Written Communication* 32/1: 39–65.
Daintith, J., and E. Martin (eds.). (2010). *Oxford Dictionary of Science*. Oxford: Oxford University Press.
Darian, S. (2003). *Understanding the Language of Science*. Austin: University of Texas Press.
Dawson, P. (2015). "Ten theses against fictionality." *Narrative* 23/1: 74–100.
De Oliveira, J., and A. Pagano. (2006). "The research article and the science popularization article: A probabilistic functional grammar perspective on direct discourse presentation." *Discourse Studies* 8/5: 627–646.
Deslauriers, M. (2007). *Aristotle on Definition*. Boston: Brill.
Doudna, J. (2017). "A crack in creation: Gene editing and the unthinkable power to control evolution." Aspen Ideas Festival 26 June. Aspen, Colorado: The Aspen Institute.
Eggins, S., and D. Slade. (1997). *Analyzing Casual Conversation*. London: Cassell.
Fairclough, N. (1989/2013). *Language and Power*. London: Routledge.
Fine, A. (1993). "Fictionalism." *Midwest Studies in Philosophy* 18: 1–18.
"First human-pig embryos made, then destroyed, as part of research into transplantable organs." (2017). *CNN Wire*. 26 Jan. Accessible: http://ktla.com/2017/01/26/first-human-pig-embryos-made-then-destroyed/ Accessed: 6 May 2018.
Fleischman, S. (1997). "The 'Labovian model' revisited with special consideration of literary narrative." *Journal of Narrative and Life History* 7/1–4: 159–168.
Fludernik, M. (1996). *Towards a "Natural" Narratology*. New York: Routledge.
Fu, X., and K. Hyland. (2014). "Interaction in two journalistic genres: A study of interactional metadiscourse." *English Text Construction* 7/1: 122–144.
Genecov, M. (2018). "What Stephen Hawking's final paper really means." *The Outline*. 13 May. Accessible: https://theoutline.com/post/4535/stephen-hawking-final-paper-normal-science?zd=1&zi=bk4wlxvi Accessed: 13 May 2018.
Gensler, H. J. (2002) *Introduction to Logic*. London: Routledge.
Gilbert, G., and M. Mulkay. (1984). *Opening Pandora's Box: A Sociological Analysis of Scientists' Discourse*. Cambridge: Cambridge University Press.
Gordin, M. D. (2015). *Scientific Babel: How Science Was Done Before and After Global English*. Chicago: University of Chicago Press.
Gülich, E. (2003). "Conversational techniques used in transferring knowledge between medical experts and non-experts." *Discourse Studies* 5/2: 235–263.

Harré, R. (1994). "Some narrative conventions of scientific discourse" in Nash, C. (ed.), *Narrative in Culture: The Uses of Storytelling in Sciences, Philosophy, and Literature*, 81–101. London: Routledge.

Harris, R., and C. Hutton. (2007). *Definition in Theory and Practice: Language, Lexicography, and the Law.* London: Continuum.

Hawking, S. (2017). "Foreword." *The Illustrated A Brief History of Time.* New York: Bantam Books.

Herman, D. (2009). *Basic Elements of Narrative.* Hoboken: Wiley-Blackwell.

Hermwille, L. (2016). "The role of narratives in socio-technical transitions—Fukushima and the energy regimes of Japan, Germany, and the United Kingdom." *Energy Research and Social Science* 11: 237–246.

Hoey, M. (1983). *On the Surface of Discourse.* London: George Allen and Unwin.

Hoey, M. (2001). *Textual Interaction: An Introduction to Written Discourse Analysis.* London: Routledge.

Hyland, K. (2005a). *Metadiscourse: Exploring Interaction in Writing.* London: Continuum.

Hyland, K. (2005b). "Representing readers in writing: Student and expert practices." *Linguistics and Education* 16: 363–377.

Hyland, K. (2010). "Constructing proximity: Relating to readers in popular and professional science." *Journal of English for Academic Purposes* 9: 116–127.

Hyvärinen, M. (2010). "Revisiting the narrative turns." *Life Writing* 7/1: 69–82.

Iser, W. (1972). *The Implied Reader: Patterns of Communication in Prose Fiction from Bunyan to Beckett.* Baltimore: John Hopkins University Press.

Keene, M. (2014). "Familiar science in nineteenth-century Britain." *History of Science,* 52/1: 53–71.

Koteyko, N., B. Nerlich, P. Crawford, and N. Wright. (2008). "'Not rocket science' or 'no silver bullet'? Media and government discourses about MRSA and cleanliness." *Applied Linguistics* 29/2: 223–243.

Kranich, S. (2009). "Epistemic modality in English popular scientific texts and their German translations." *Journal of Translation and Technical Communication Research* 2/1: 26–41.

Kranich, S. (2011). "To hedge or not to hedge: The use of epistemic modal expressions in popular science in English texts, English-German translations, and German original texts." *Text and Talk* 31/1: 77–99.

Kranich, S. (2016). *Contrastive Pragmatics and Translation. Evaluation, Epistemic Modality and Communicative Style in English and German.* Amsterdam: John Benjamins.

Kranich, S., J. House, and V. Becher. (2012). "Changing conventions in English-German translations of popular scientific texts" In Braunmüller, K., and C. Gabriel (eds.), *Multilingual Individuals and Multilingual Societies*, 315–334. Amsterdam: John Benjamins.

Kreiswirth, M. (1992). "Trusting the tale: Narrativist turn in the human sciences." *New Literary History* 23/3: 629–657.

Labov, W. (1972). *Language in the Inner City: Studies in the Black English Vernacular.* Philadelphia: University of Pennsylvania Press.

Labov, W., and J. Waletsky. (1967). "Narrative analysis: Oral versions of personal Experience" in Helm, J. (ed.), *Essays on the Verbal and Visual Arts: Proceedings of the 1966 Annual Spring Meeting of the American Ethnological Society*, 12–44. Seattle: University of Washington Press.

Ladegaard, H. J. (1995) "Audience design revisited: Persons, roles and power relations in speech interaction." *Language and Communication* 15/1: 89–101.

Leech, G., and M. Short. (1981/2007). *Style in Fiction: A Linguistic Introduction to English Fictional Prose.* London: Pearson.

Lerer, S. *Inventing English: A Portable History of the Language.* New York: Columbia University Press, 2015.
Lewin, B., and H. Perpignan. (2012). "Recruiting the reader in literary criticism." *Text and Talk* 32/6: 751–772.
Liao, M.-H. (2010). "Translating science into Chinese: An interactive perspective." *The Journal of Specialized Translation* 13: 44–60.
Lightman, B. (2000). "Marketing knowledge for the general reader: Victorian popularizers of science." *Endeavour* 24/3: 100–106.
Livnat, Z. (2012). *Dialogue, Science and Academic Writing.* Amsterdam: John Benjamins.
Luzón, M. (2013). "Public communication of science in blogs: Recontextualizing scientific discourse for a diversified audience." *Written Communication* 30/4: 428–457.
Macedo-Rouet, M., J-F. Rouet, I. Epstein, and P. Fayard. (2003). "Effects of online reading on popular science comprehension." *Science Communication* 25/2: 99–128.
Mair, C. (2006). *Twentieth-Century English: History, Variation, and Standardization.* Cambridge: Cambridge University Press.
Martin, J.R., and P.R.R. White. (2005). *The language of Evaluation: Appraisal in English.* New York: Palgrave Macmillan.
McCarthy, M. (2005). *Discourse analysis for language teachers.* Cambridge: Cambridge University Press.
Medawar, P. B. (1984). *The Limits of Science.* New York: Harper & Row.
Mellor, F. (2003). "Between fact and fiction: Demarcating science from non-science in popular physics books." *Social Studies of Science* 33/4: 509–538.
Meredith, D. (2010) *Explaining Research: How to Reach Key Audiences to Advance Your Work.* Oxford: Oxford University Press.
Merton, R. K., and E. Barber. (2004). *The Travels and Adventures of Serendipity: A Study in Sociological Semantics and the Sociology of Science.* Princeton: Princeton University Press.
Mildorf, J. (2008). "Thought presentation and constructed dialogue in oral stories: Limits and possibilities of a cross-disciplinary narratology." *Partial Answers: Journal of Literature and the History of Ideas* 6/2: 279–300.
Millar, N., B. Budgell, and K. Fuller. (2013). "'Use the active voice whenever possible': The impact of style guidelines in medical journals." *Applied Linguistics* 34/4: 393–414.
Moirand, S. (2003). "Communicative and cognitive dimensions of discourse on science in the French mass media." *Discourse Studies* 5/2: 175–206.
Moon, R. (2009). "Sinclair, lexicography, and the Cobuild project: The application of theory" in Rosamund Moon (ed.), *Words, Grammar, Text: Revisiting the Works of John Sinclair,* 1–22. Amsterdam: John Benjamins.
Mugglestone, L. (2000) *Lexicography and the OED: Pioneers in the Untrodden Forest.* New York: Oxford University Press.
Myers, G. (1990) *Writing Biology: Texts in the Social Construction of Scientific Knowledge.* Madison: University of Wisconsin Press.
Myers, G. (1992). "Fictions for facts: The form and authority of the scientific dialogue." *History of Science* 30: 221–247.
Myers, G. (1997). "Fictionality, demonstration, and a forum for popular science: Jane Marcet's *Conversations on Chemistry*" in Gates, B. T., and A. B. Shteir (eds.), *Natural Eloquence: Women Reinscribe Science,* 43–60. Madison: University of Wisconsin Press.
Myers, G. (1999). "Functions of reported speech in group discussions." *Applied Linguistics* 20/3: 376–401.
Myers, G. (2003). "Discourse studies of scientific popularisation: Questioning the boundaries." *Discourse Studies* 5/2: 265–279.
Myers, G. (2010). *Discourse of blogs and wikis.* London: Continuum International.

Nicolopoulou A. (1997). "Labov's legacy for narrative research—and its ironies." *Journal of Narrative and Life History* 7/1–4: 369–377.

Olson, R. (2015). *Houston, We Have a Narrative: Why Science Needs Story.* Chicago: University of Chicago Press.

Parkinson, J., and R. Adendorff. (2004). "The use of popular science articles in teaching scientific literacy." *English for Specific Purposes* 23: 379–396.

Peterson, C., and A. McCabe. (1983). *Developmental Psycholinguistics: Three Ways of Looking at a Child's Narrative.* London: Plenum Press.

Pilkington, O. (2010). "Americanization of Russian culture and its effects on English language acquisition in that country." *Journal of the Utah Academy of Sciences, Arts & Letters* 87: 283–292.

Pilkington, O. (2017). "Popular science versus lab lit: Differently depicting scientific apparatus." *Science as Culture* 26/3: 285–306

Pilkington, O. (2018). *Presented Discourse in Popular Science: Professional Voices in Books for Lay Audiences.* Leiden, Netherlands: Brill.

Pilkington, O. (forthcoming). "Uhura and the linguistics of *Star Trek*" in Kapell, M. (ed.), *Star Trek: Kelvin Timeline.* Jefferson, North Carolina: McFarland.

Pilkington, O., and A. Pilkington (eds.) (forthcoming). *Lab Lit: Exploring Literary Fictions about Science.* Lanham, Maryland: Lexington.

Plato. (1956). "Meno" in Warmington, E. H., and P. G. Rouse (eds.), W. H. D. Rouse (transl.), *Great Dialogues of Plato*, 26–68. New York: New American Library.

Plato. (1961). "Cratylus" in Hamilton, E., and H. Cairns (eds.), B. Jowett (transl.), *The Collected Dialogues of Plato, Including the Letters*, 421–474. New York: Pantheon Books.

Plato. (2011). "Euthyphro" in Cohen, S. M., P. Curd, and C. D. C. Reeve (eds.), *Readings in Ancient Greek Philosophy: From Thales to Aristotle*, 135–152. Indianapolis: Hackett.

Polanyi, L. (1979). "So what's the point?" *Semiotica* 25/3–4: 207–242.

Prashadjan, V. (2018). "How a church deacon found the biggest prime number yet (it wasn't as hard as you think)." *The New York Times.* 26 Jan. Accessible: https://www.nytimes.com/2018/01/26/science/prime-number-mersenne-church.html Accessed: 29 May 2018.

Reitsma, F. (2010). "Geoscience explanations: Identifying what is needed for generating scientific narratives from data models." *Environmental Modelling & Software* 25: 93–99.

Roberts, K. (2013). "Foreword" in Bowater, L., and K. Yeoman, *Science Communication: A Practical Guide for Scientists*, xix. Chichester: Wiley-Blackwell.

Robinson, R. (1962). *Definition.* Oxford: Clarendon Press.

Rohn, J. L. (2005). "What is lab lit (the genre)?" Accessible: http://www.lablit.com/article/3. Accessed: 5 March 2015.

Rohn, J. L. (2010a). "More lab in the library." *Nature* 465/3: 552.

Rosenthal, D. (1993) "Images of scientists: A comparison of biology and liberal studies majors." *School Science and Mathematics* 93/4: 212–216.

Rouse, J. (2009) "Laboratory fictions" in Suarez, M. (ed.), *Fictions in Science: Philosophical Essays on Modeling and Idealization*, 37–55. London: Routledge.

Sackler, M. (ed.). (2014). *Science of Science Communication II: Summary of a Colloquium.* Washington, D.C.: The National Academies Press.

Santas, G. X. (1999). *Socrates: The Arguments of the Philosophers.* New York: Routledge.

Schank, R. C. (1990). *Tell Me a Story: A New Look at Real and Artificial Memory.* New York: Atheneum.

Schiappa, E. (2003). *Defining Reality: Definitions and the Politics of Meaning.* Carbondale: Southern Illinois University Press.

Schrödinger, E. (1935). "Die gegenwärtige Situation in der Quantenmechanik (The present

situation in quantum mechanics)." *Naturwissenschaften* 23/48: 807–812. Accessible: https://link.springer.com/article/10.1007%2FBF01491891 Accessed: 5 June 2018.
Semino, E., and M. Short. (2004). *Corpus Stylistics: Speech, Writing and Thought Presentation in a Corpus of English Writing.* London: Routledge.
Short, M. (2007). "Thought presentation twenty-five years on." *Style* 41/2: 225–241.
Short, M. (2012). "Discourse presentation and speech (and writing but not thought) summary." *Language and Literature* 21/1: 18–32.
Short, M., E. Semino, and M. Wynne. (2002). "Revisiting the notion of faithfulness in discourse presentation using a corpus approach." *Language and Literature* 11/4: 325–355.
Skov Nielsen, H., J. Phelan, and R. Walsh. (2015a). "Ten theses about fictionality." *Narrative* 23/1: 61–73.
Skov Nielsen, H., J. Phelan, and R. Walsh. (2015b). "Fictionality as rhetoric: A response to Paul Dawson." *Narrative* 23/1: 101–111.
Smirnova, A.V. (2009). "Reported speech as an element of argumentative newspaper discourse." *Discourse & Communication* 3/1: 79–103.
Smolin, L. (2007). *The Trouble with Physics: The Rise of String Theory, the Fall of a Science, and What Comes Next.* New York: Houghton Mifflin.
Souba, W. (2011). "The language of discovery." *Journal of Biomedical Discovery and Collaboration* 6: 53–69.
Suarez, M. (2009). "Fictions in scientific practice" in Suarez, M. (ed.), *Fictions in Science: Philosophical Essays on Modeling and Idealization*, 3–15. London: Routledge.
Supper, A. (2014). "Sublime frequencies: the construction of sublime listening experiences in the sonification of scientific data." *Social Studies of Science* 44/1: 34–58.
Sydserff, R., and P. Weetman. (1999). "A texture index for evaluating accounting narratives: An alternative to readability formulas." *Accounting, Auditing, and Accountability Journal* 12: 459–458.
Tagg, C. (2009). *A Corpus Linguistics Study of sms Text Messaging* (doctoral dissertation). University of Birmingham.
Talbot, M. (1995). "A synthetic sisterhood: False friends in a teenage magazine" in Hall K., and M. Bucholtz (eds.), *Gender Articulated: Language and the Socially Constructed Self*, 143–168. New York: Routledge.
Tannen, D. (2007). *Talking Voices: Repetition, Dialogue, and Imagery in Conversational Discourse.* Cambridge: Cambridge University Press.
Tappan, M. B. (1997). "Analyzing stories of moral experience: narrative, voice, and the dialogical self." *Journal of Narrative and Life History* 7/1–4: 379–386.
Thompson, G. (2012). "Intersubjectivity in newspaper editorials: Constructing the reader-in-the-text." *English Text Construction* 5/1: 77–100.
Thompson, G., and P. Thetela. (1995). "The sound of one hand clapping: The management of interaction in written discourse." *Text* 15/1: 103–127.
Thompson G. (2001) "Interaction in academic writing: Learning to argue with the Reader." *Applied Linguistics* 22/1: 58–78.
Tlauka, M., and F. P. McKenna. (1998). "Mental imagery yields stimulus-response compatibility." *Acta Psychologica* 98: 67–79.
Toolan, M. (1996). *Total Speech: An Integrational Linguistic Approach to Language.* Durham: Duke University Press.
Toolan, M. (2001). *Narrative: A Critical Linguistic Introduction.* London: Routledge.
Toolan, M. (2011). "The texture of emotionally-immersive passages in short stories: Steps towards a tentative local grammar" in *Proceedings of the 3rd International Stylistics Conference.* Accessible: http://artsweb.bham.ac.uk/MToolan/UPLOADS/MICHAEL%20TOOLAN%20Ningbo%20March%202011%20-%20Local%20Grammar.pdf Accessed: 16 September 2014.

Toon, A. (2012). *Models and Make-Believe: Imagination, Fiction and Scientific Representation*. New York: Palgrave McMillan.

Topham, J. (2000). "Scientific publishing and the reading of science in nineteenth-century Britain: A historiographical survey and guide to sources." *Studies in History and Philosophy of Science* 31/4: 559–612.

Turney, J. (2004a). "The abstract sublime: Life as information waiting to be rewritten." *Science as Culture* 13/1: 89–103.

Turney, J. (2004b). "Accounting for explanation in popular science texts—an analysis of popularized accounts of superstring theory." *Public Understanding of Science* 13: 331–346.

Turney, J. (2007). "Boom and bust in popular science." *Journal of Science Communication* 6/1: 1–3.

Urbanova, Z. (2012). "Direct and free direct forms of representation in the discourse of newspaper reports: Less frequent phenomena." *Brno Studies in English* 38/1: 39–54.

Vaihinger, H. (1924/2001). *The Philosophy of "As If"* C. K. Ogden (trans.). London: Routledge.

Varttala, T. (1999). "Remarks on the communicative functions of hedging in popular scientific and specialist research articles on medicine." *English for Specific Purposes* 18/2: 177–200.

Verschueren, J. (2004). "Notes on the role of metapragmatic awareness in language use" in Jaworski, A., N. Coupland, and D. Galasinski (eds.), *Metalanguage: Social and Ideological Perspectives*, 53–74. Berlin: Mouton de Gruyter.

Waugh, L. (1995). "Reported speech in journalistic discourse: The relation of function and text." *Text* 15/1: 129–173.

White, M., and J. Gribbin. (1992). *Stephen Hawking: A Life in Science*. New York: Viking.

Zarkadakis, G. (2016). *In Our Own Image: Savior or Destroyer? The History and Future of Artificial Intelligence.* New York: Pegasus Books.

Index

affect 135, 142; *see also* emotionality
Aristotle 29–32, 126, 127, 130, 149
audience design 39, 40, 135–137, 146

Bryson, Bill 2, 14, 42, 45–46, 52, 100, 104–105, 111–112, 145
Bynum, William 27, 42, 109, 118

Carroll, Sean 42, 98–99, 119–120, 153
chimera science 9
clause, reporting 25–27, 94, 101, 103, 108, 110; reported 101
Coen, Enrico 15, 42, 55–56, 67

Davy, Humphry 70, 167
definition 2, 9, 11, 29–41, 62, 122–149, 170, 175
dialogue 21, 25, 30, 92–99, 105, 152, 170, 176–177
discourse 41, 43, 122, 151, 170, 177; celebratory 10–11, 89–90, 159, 162 (*see also* science, positive presentation of); contingent 85; empiricist 86; presented 2, 11, 14, 21, 24–29, 91–121, 131, 144, 152–155, 169; scientific 7, 60, 68, 71, 85, 147, 173
discovery 8, 12, 14, 71, 100–101, 116–118, 159, 173, 175–176; presentation of 99, 107–108, 110, 112–113; *see also* narrative of
displacement: spatial 20; temporal 20
dramatization 21, 91–96, 98, 100, 105, 113–114, 150, 169; *see also* emotionality; fictionality
du Sautoy, Marcus 2, 13, 37, 42, 60–64, 68, 75, 77, 80, 86–87, 101, 123, 128, 130, 132–133, 138–140, 142, 145–148, 154, 176

emotionality 94–98, 100–102, 167
evaluation (narrative element) 13, 44, 48–49, 53, 59–62, 65–67, 70, 72–74, 79–80; waves of 49, 62
experientiality 91
explanation (narrative element) 13–14, 50, 59–68; waves of 62–63

Ferris, Timothy 2, 42, 100, 102, 104–105, 109, 113–114, 119
fiction about scientists 2; *see also* lab lit
fictionality 9, 20–21, 93, 101, 113, 166–171, 175; *see also* dramatization; emotionality; reader, fictionalized

Gaines, Susan 2, 158, 161
Goodman, Allegra 2, 95, 158, 160–161
Greene, Brian 2, 9, 12, 15, 35–37, 39–40, 42, 46–51, 53, 55, 57, 64–65, 67–68, 72–74, 78, 81–84, 88, 102–103, 112, 116, 124, 127–133, 137, 141, 143–146, 148, 152–154, 157–159, 165

Hawking, Stephen 1–2, 18, 74, 112–113
Hoey, Michael 51–54, 68–69, 71–74, 80–81, 83, 89
Huxley, T.H. 5–6, 122
hypotheses: negative 111, 114; positive 110, 114; presentation of 97, 107–110; verb patterns 108

Kaku, Michio 2, 10, 41, 42, 46, 53–55, 66, 76, 80, 99, 124–128, 141, 144–146, 148, 153, 155, 165

187

Index

Kean, Sam 2, 15, 25, 27, 42, 95–96, 99, 102, 111, 117–118

lab lit 2, 156–163
Labov, William 13, 44–45, 48–51, 53–55, 59–62, 65–66, 68–69
language: English 29, 36, 42, 45, 51, 137, 151, 172–177; figurative 39, 102–105, 111, 113–114, 123, 131, 137, 142–143, 146, 149
lexical signal 53, 71–73, 75, 79, 82
Lightman, Alan 2, 158, 162–163

McEwan, Ian 157

narrative 2, 8–9, 12–13, 18–24, 41–43, 129, 152, 154, 163, 166, 168–171; chance 86–89; conflict 72, 81–86; of discovery 11, 13–14, 43, 45, 50, 52, 54–55, 57, 59–69, 70–90, 99, 109, 111, 120–121; elements of 23, 44, 48, 50–51, 53–56, 59, 62–63, 65–66, 68; of failed discoveries 55, 72, 79, 81, 86, 111; macrostructure 44–45, 65, 68; microstructure 44–45; narrative turn 22; personal 44–58, 67; skeleton 45

Pepper, Henry 70, 167
power structure 8, 146, 148; initiating 146–147, 149; responsive 146

reader: engagement 139, 155; fictionalized 150–155; reader-in-the-text 137, 151, 153
Rohn, Jennifer 2, 156–158, 160–162

scaffolding 13, 59–60, 64–66, 140; cognitive load 144–145
science, positive presentation of 9–10, 23, 55–57, 71, 75, 80, 87, 89, 101, 104, 108, 141–143, 146, 159, 162, 165; *see also* discourse, celebratory; hypotheses, positive
self-categorization 146
Socrates 30–31, 33, 127, 149
solidarity 135, 137, 147–148; exclusive 148; inclusive 148
sonification 10

temporal sequence 22, 44–45, 49–51, 60–63, 66; *see also* narrative
theatricality 70, 167

writing: popular scientific 172, 174; presentation of 115–121; professional scientific 5, 176

Zimmer, Carl 8, 26, 42, 109, 110, 145

www.ingramcontent.com/pod-product-compliance
Lightning Source LLC
Chambersburg PA
CBHW032102300426

44116CB00007B/858